Praise for Dhaval Bathia

Dhaval Bathia packs in quite a cerebral punch. A unique blend of talent, insight, hard work and sheer determination.

— *Education Times, The Times of India*

A WHIZ KID. Dhaval Bathia needs just a split second to recall details of 100 year old calendars, hundreds of phone numbers and to solve complex arithmetical problems.

— *Bombay Times, The Times of India*

A unique achiever. We wish him all the best for his future projects.

— *Lohana Shakti*

Dhaval Bathia is simply 'excellent'. His workshops are commendable.

— *Maharashtra Times*

By using simple word formulae he manages to solve complex arithmetical calculations within seconds.

— *Janmabhoomi*

A young achiever. His seminars receive a tremendous response even from the corporate world.

— *Education World*

Dhaval Bathia is the king of Vedic Mathematics

— Samakaleen

White-Lightning success in Vedic Mathematics

— Sakal

He can recollect all facts of any subject by reading them just once!

The manner in which the student community has purchased his book *How To Top Exams and Enjoy Studies* might well assure him a career in counseling.

— Navbharat Times

Mathematical genius

— Kuwait Times

After attending his seminars, students have found a sharp increase in their confidence level and their performance in exams has also improved.

— Abhiyaan

Dhaval Bathia's seminar was definitely a high point. He tantalized the crowd with his mathematical prowess.

— JAM Magazines

This youngster has mastered Mathematics the Vedic way.

— The Indian Express

He solves problems on addition, subtraction, multiplication, division, roots and any other such problems instantly.

— Gujarat Samachar

The scientific systems in the book (*How to Top Exams & Enjoy Studies*) create a paradigm shift from hard work to smart work. The whole emphasis is laid on the fact that education is a process to be enjoyed and cherished.

— Afternoon Despatch & Courier

Indian Wizard

— The Gulf Today (UAE)

Dhaval's number tricks are unbelievable…he can tell you when is your birthday without you telling him!

— Le Mauricien (Mauritius)

He kept the audience glued, mesmerized, with his techniques

— Al Bayan (UAE)

Our country not only produces great talent but also nurtures it so that it reaches its fullest potential which enables it to reach the sky. Since time immemorial India has not only given great intellectuals, sportspersons, revolutionaries, socialists but also great artists, painters, scientists to the countrymen and society in general. Adding another name to this great stardom is such an amazing talent born in the Bathia family of Maharashtra. Shri Dhaval Bathia has researched on some unique and exciting techniques of Vedic Mathematics…

— Jazbaat Magazine

An inspiration to the youth

— FM Gold

Dhaval Bathia is a revolution…

— S TV

Vedic Mathematics

Made Easy

First book on Vedic Math with VCD

DHAVAL BATHIA'S BOOKS HAVE SOLD 1.5 LAKH COPIES WORLDWIDE
COPIES 150,000 SOLD

Dhaval Bathia

JAICO PUBLISHING HOUSE

Ahmedabad Bangalore Bhopal Bhubaneswar Chennai
Delhi Hyderabad Kolkata Lucknow Mumbai

Published by Jaico Publishing House
A-2 Jash Chambers, 7-A Sir Phirozshah Mehta Road
Fort, Mumbai - 400 001
jaicopub@jaicobooks.com
www.jaicobooks.com

VEDIC MATHEMATICS MADE EASY
With VCD
ISBN 978-81-8495-301-5

First Jaico Impression: 2012
Fourth Jaico Impression: 2014

Printed by
Snehesh Printers
320-A, Shah & Nahar Ind. Est. A-1
Lower Parel, Mumbai - 400 013.

Dedicated
to
Jay

Contents

Acknowledgments

And there are those who give and know not pain in giving,
nor do they seek joy, nor give with mindfulness of virtue;
They give as in yonder valley the myrtle breathes its
fragrance into space
Through the hands of such as these God speaks and from
behind their eyes
HE smiles upon the earth

— KAHLIL GIBRAN

My sincerest gratitude to my parents, family members, friends and relatives for being my pillar of strength...

To all participants of my seminars for turning out in huge numbers every time I conduct a workshop.

To my readers for their constant suggestions and constructive criticisms for the improvement of my books...

About the Author

Dhaval Bathia (B.Com, ACS, LLB) is one of the world's youngest international bestselling authors. He has four books to his credit which have been translated in more than 14 languages.

A pioneer in the field of Vedic Mathematics, he gave his first seminar on Vedic Mathematics at the age of 16. At seventeen, he started teaching Vedic Mathematics and other mind-power sciences to professors in various educational institutions and urged them to spread his systems amongst the student community. He has trained over 400,000 students on the subject in India, Kuwait, UAE, USA, Mauritius, Nepal, Oman and UK.

E-Mail: dhaval@dhavalbathia.com

Website: www.dhavalbathia.com

Twitter: www.twitter.com/dhavalbathia

Blog: http://extremelyconfidential.blogspot.com

Facebook Community: 'Dhaval Bathia's Community Page

(To join Vedic Maths e-group and share and receive techniques, send a mail to vedic_mathematics-subscribe@yahoogroups.com)

Introduction

You must have heard of Vedic Mathematics but wondered what this was all about.

Vedic Mathematics is the collective name given to a set of sixteen mathematical formulae as discovered by Jagadguru Swami Sri Bharati Krishna Tirthaji Maharaj. Each formula deals with a different branch of Mathematics. Complex mathematical questions which otherwise take numerous steps to solve can be solved with the help of a few steps and in some cases without any intermediate steps at all! And these systems are so simple that even people with an average knowledge of mathematics can easily understand them.

Four Exciting Magic Tricks

Every second person whom I meet tells me that he hates mathematics. As they hate mathematics, they get poor scores in their exams and the poor scores further aggravate their hatred. Thus it becomes a vicious circle!

Over the years I realized that people often came to my seminars with a negative mindset towards mathematics. Since an early age, their mind is conditioned to believe that maths is boring and difficult. Thus, it becomes inevitable to create a 'paradigm shift' towards the subject. Unless they are excited about the techniques that will follow, it becomes a Herculean task to arouse enthusiasm towards the seminar.

With a desire to create a curiosity and yearning for the subject, we researched on a set of techniques which we call 'Mental Magic.' These techniques are so exciting and powerful that they leave the audience bewildered. They get tantalized with its workings and their entire attitude towards mathematics changes in a few minutes. After the session on Mental Magic is over, the audience is all eager, excited and ready to grasp the techniques of Vedic Mathematics. This really simplifies my task.

In this chapter, we are going to study many of these techniques which will change your attitude towards mathematics. A few of these are trade secrets, but for the first time I am revealing them in my book.

In such sessions, I predict the date of birth of anybody from the audience. I also predict strange things like how many brothers/sisters he has, how many children he has, how much money one has in his pockets etc., without him disclosing any information. I also do a technique where I can get the answer to a problem without even knowing the problem.

Sounds interesting! Read ahead...

(A) How to predict a person's Date of Birth

With this technique you can predict the date of birth of any number of people simultaneously. You can try this stunt with your family members, friends, relatives, colleagues and even in parties.

STEPS

    (a) Ask the people to take the number of the month in which they were born (January is 1, February is 2 and so on.....)

    (b) Next, ask them to double the number

    (c) Add 5 to it

    (d) Multiply it by 5

    (e) Put a zero behind the answer

    (f) Add their date of birth (If they are born on 5th January then add 5)

After the steps are over, ask them to tell you the final answer. And lo! Just by listening to their final answer you can predict their date of birth !

SECRET

From the answer that you get from each member of the audience,

- Mentally subtract 50 from the last two digits and you will have the date
- Subtract 2 from the remaining digits and you will have the month.

Thus, you will easily get his date of birth.

Example:

Let us suppose I was a member of the audience. My date of birth is 26th June. Then I would have worked out the steps as follows:

(a)	Take the month number	=	6
(b)	Double the answer	=	12
(c)	Add 5 to it	=	17
(d)	Multiply it by 5	=	85
(e)	Put a zero behind the answer	=	850
(f)	Add the date of birth	=	850 + 26 = 876

Thus my final answer is 876. Now, let us see how we can deduct my date of birth from the final answer. We will subtract 50 from the last two digits to get the date. Next, we will subtract 2 from the remaining digits to get the month.

$$
\begin{array}{r|r}
8 & 76 \\
-\ 2 & 50 \\
\hline
6 & 26 \\
\end{array}
$$
(month) | (date)

On similar lines if the total was 765, 1480 and 1071 the birthdates would be 15th May, 30th December and 21st August respectively.

$$
\begin{array}{r|r}
7 & 65 \\
-\ 2 & 50 \\
\hline
5 & 15 \\
\end{array}
\qquad
\begin{array}{r|r}
14 & 81 \\
-\ 2 & 50 \\
\hline
12 & 31 \\
\end{array}
\qquad
\begin{array}{r|r}
10 & 71 \\
-\ 2 & 50 \\
\hline
8 & 21 \\
\end{array}
$$

With the knowledge of this technique, you can predict the birthdate of hundreds of people simultaneously.

(Note: There are many such mathematical ways by which you can predict a person's date of birth but the method given above is one of the simplest methods)

(B) How to predict how much money a person has in his pocket

This technique will help you find how much money a person has in his wallet/pocket etc. It can be tried on a group of people simultaneously.

STEPS

  (a) Ask him to take the amount he has in his pocket (just the rupees, ignore the paisa)

  (b) Next ask him to add 5 to it

  (c) Multiply the answer by 5

  (d) Double the answer so obtained

  (e) Finally, ask him to add his favourite one digit number (any number from 0 to 9)

  (f)   Add 10

After the steps are over ask them to tell you the final answer. And just by listening to the final answer you will come to know the amount he has in his pocket!

SECRET

  • Ignore the digit in the unit's place.
  • From the remaining number, subtract 6 and you will come to know the amount he has in this pocket.

Example:

Let us suppose a person has 20 Rupees in his pocket. He would work out the steps as given below:

(a) Take the amount in your pocket	=	20
(b) Add 5	=	25
(c) Multiply by 5	=	125
(d) Double the answer	=	250
(e) Add your favourite single digit number (say, 7)	=	257
(f) Add 10	=	267

Thus his final answer would have been 267. Now let us see how we can find out the amount he has from the final answer. As mentioned earlier, we will ignore the digit in the unit's place (in this case it is 7). Now the remaining number is 26. From 26, we subtract 6 to get 20. Thus our answer is confirmed.

Similarly if the total was 1062, 63 and 170 the amount would be 100, 0 and 11 respectively.

(From 1062, we ignore the last digit 2 and take only 106. From 106, we subtract 6 and get the answer as 100 and so on...)

I use this technique in my seminars to give the audience some relief from complex calculations. There are many ways by which we can predict the amount a person has in his pocket but I prefer this technique because it neither involves any complex calculations nor is it too obvious for the audience to guess the secret.

(C) How to find out how many brothers/sisters a person has

We have seen how to find out a person's birthday and how to find out how much money a person has in his pocket. Now, we will see how to find out how many brothers and sisters a person has. As always, you can try this trick on others to impress them. However, when a person is doing these calculations ask him to take only his siblings (children of the same parents) and ignore his cousins.

STEPS

(a) Ask the person, (on whom you are trying the trick) to take the number of brothers he has (if no brother; take zero)

(b) Add 3 to it

(c) Multiply it by 5

(d) Add 20

(e) Double the answer

(f) Now, ask him to add the number of sister he has (if no sisters, take zero)

(g) Finally, ask him to add 1

After the steps are over ask him to reveal the final answer. And as unbelievable it might sound, within a second you will be able to guess how many brothers and sisters he has!

SECRET

- From the final answer that the other person will tell you, mentally subtract 71

- The last digit will give you number of sisters and the remaining digits will give you the number of brothers he has.

Example:

Let us suppose a person has 1 brother and 1 sister. This is how he would work out the steps.

(a) Take the number of brothers =   1

(b) Add 3 to it   = 4

(c) Multiply by 5   = 20

(d) Add 20   = 40

(e) Double the answer   = 80

(f) Add number of sisters   = 81 (80 plus 1 sister = 81)

(g) Add 1   = 82

Thus, his final answer would be 82. Now, let us see how we can find out his siblings from the final answer. As mentioned earlier, we will subtract 1 from the last digit to get the number of sisters. Next, we will subtract 7 from the first digit to get the number of brothers.

$$
\begin{array}{c|c}
8 & 2 \\
- \ 7 & 1 \\
\hline
1 & 1 \\
\text{(brother)} & \text{(sister)}
\end{array}
$$

Similarly if the total was 71, 93, 102 the number of brothers and sisters would be (0, 0), (2, 2) and (3, 1) respectively.

$$
\begin{array}{c|c} 7 & 1 \\ - \ 7 & 1 \\ \hline 0 & 0 \end{array}
\qquad
\begin{array}{c|c} 9 & 3 \\ 7 & 1 \\ \hline 2 & 2 \end{array}
\qquad
\begin{array}{c|c} 10 & 2 \\ 7 & 1 \\ \hline 3 & 1 \end{array}
$$

(Similarly, you can predict a how many sons and daughters a person has by substituting the word 'brother' with 'sons' and 'sisters' with 'daughters' in the above examples)

Now let's gear up for the last magic. This is slightly different, but very exciting!

(D) How to find the answer without knowing the question !

Using this technique you can find the total of a set of five numbers without knowing the numbers.

In my seminars, I ask a member of the audience to give me a three-digit number. Let us suppose someone gives me the number 801. Then, I write the number 801 and after leaving four lines I write the final answer as 2799.

801 (Audience)

Ans: 2799

Thus, I have got the final answer without knowing all the numbers. Next, I ask him to give me another three digit number. Let us suppose he gives me 354. Now, its my turn. Within one second, I write the next number as 645.

801 (Audience)
354 (Audience)
645 (Speaker)

Ans: 2799

Now, again it's the turn of the audience. Let us suppose he gives me 800. Now, again it is my turn. Before he blinks his eye, I write my number as 199.

801 (Audience)
354 (Audience)
645 (Speaker)
800 (Audience)
199 (Speaker)

Ans: 2799

Thus, I have all five numbers in place. When I ask him to check the total of the five numbers, he is amazed to find out that it is 2799 which I had predicted by looking at the first number only !

Thus, I got the answer without knowing the question. Further, after every step I was writing my three-digit number within one second (without any time for calculation)

Even you can do such 'Magic Totals.' You can ask the audience to give you any three-digit number and based on that you can get the final total without looking at the other numbers.

SECRET

- From the first number that the audience gives you, **subtract 2 and always put 2 in the beginning**. This becomes your final answer. For example, if the number is 801, we subtract 2 and get 799. Next, we put 2 in the beginning and our final answer becomes 2799.

  If the number is 567, we subtract 2 and get 565. Next, we put 2 in the beginning and write the final answer as 2565. Thus, you will have the final answer.

- Next, we will see the secret of finding the subsequent steps. **Subtract each digit of the number that the audience gives you from 9** and you will have your number. In the example mentioned above, when the audience gives me 354, I subtract each of the digits 3, 5 and 4 from 9 and get my answer as 6, 4 and 5. So my number is 645. When the audience gives me the number 800, I subtract each of the digits 8, 0 and 0 from 9 and get my answer as 1, 9 and 9. So my number is 199.

In this way, you can have the intermediary steps.

Let us have a look at another example.

600 (A)	600 (A)	600 (A)
→ 481 (A)	→ 481 (A)
	518 (S)	518 (S)
		909 (A)
		090 (S)
Ans: **2598**	Ans: **2598**	Ans: **2598**

(A = Audience, S = Speaker)

If the audience gives you a number as 600, you will subtract 2 from it (598) and put 2 in the beginning. Thus, your final answer is 2598. Next, let us assume the audience gives you the number 481. We subtract each of the digits 4, 8 and 1 from 9 and get our answer as 518 instantly. Next, let us suppose the audience gives you the number 909. We subtract each of the digits 9, 0 and 9 from 9 and get our answer as 090. When you check the final total, you will be surprised to find that it is indeed 2598 !

Thus, I have taught you four exciting techniques of mathematics which you can use to impress your friends and colleagues. The intention of including this chapter in the book was twofold. First, since most people have got habituated to the calculator, it is a good opportunity to tune your mind to mental calculation. Secondly, the wonderful response that you will get by using these techniques on people will definitely change your attitude towards mathematics.

BASIC LEVEL

CHAPTER 1

Miscellaneous Simple Method

In Vedic Mathematics, there are two types of techniques: specific techniques and general techniques. The specific techniques are those which are fast and effective but can be applied only to a particular combination of numbers. For example, the technique of squaring numbers ending with 5 is a specific technique because it can be used to square only those numbers that end with 5. It cannot be used to square any other type of number. On the other hand, the technique of multiplication as given by the Criss-Cross System is a general technique, as it can be used to multiply numbers of any possible combination of digits.

Thus, general techniques have a much wider scope of application than specific techniques because they deal with a wider range of numbers. In this book, we will give more emphasis to general techniques as they provide a much wider utility. Chapter 1 deals with the specific techniques; from Chapter 2 onwards we will study general techniques.

We will discuss the following techniques in this chapter:

(a) Squaring of numbers ending with 5

(b) Squaring of numbers between 50-60

(c) Multiplication of numbers with a series of 9's

(d) Multiplication of numbers with a series of 1's

(e) Multiplication of numbers with similar digits in the multiplier

(f) Subtraction using the rule 'All from 9 and the last from 10'

The first technique that we will discuss is how to instantly square numbers whose last digit is 5. Remember, squaring is multiplying a number by itself. When we multiply 6 by 6 we get the answer 36. This 36 is called the square of 6.

(a) Squaring of numbers ending with '5'

Squaring is multiplying a number by itself. Let us have a look at how to square numbers ending in 5.

(Q) Find the square of 65.

$$
\begin{array}{r}
6\ 5 \\
\times\ 6\ 5 \\
\hline
42\ 25
\end{array}
$$

- In 65, the number apart from 5 is 6.
- After 6 comes 7. So, we multiply 6 by 7 and write down the answer 42.
- Next, we multiply the last digits, viz. (5 × 5) and write down 25 to the right of 42 and complete our multiplication.
- Our answer is 4225.

(Q) Find the square of 75.

$$
\begin{array}{r}
7\ 5 \\
\times\ 7\ 5 \\
\hline
56\ 25
\end{array}
$$

Apart from 5 the number is 7. The number that comes after 7 is 8. We multiply 7 with 8 and write the answer 56. Next, we multiply the last digits (5 × 5) and put 25 beside it and get our answer as 5625. Thus, (75 × 75) is 5625.

(Q) Find the square of 95.

Apart from 5 the digit is 9. After 9 comes 10. When 9 is multiplied by 10 the answer is 90. Finally, we vertically multiply the right hand most digits (5 × 5) and write the answer 25 beside it. Thus, the square of 95 is 9025.

(Q) Find the square of 105.

The previous examples that we solved were of two-digits each. But the same technique can be extended to numbers of any length. In the current example, we will try to determine the square of a three-digit number – 105.

Apart from 5 the digits are 1 and 0, that is, 10. After 10 comes 11. We multiply 10 with 11 and write the answer as 110. We suffix 25 to it and write the final answer as 11025. The square of 105 is 11025.

So you can see how simple it is to square numbers ending with a five! In fact, you can mentally calculate the square of a number ending with a 5. Just multiply the non-five numbers with the next number and then multiply the last digits (5 × 5) and add 25 after it.

A few more examples are given below:

$$15^2 = 225$$
$$25^2 = 625$$
$$35^2 = 1225$$
$$45^2 = 2025$$
$$55^2 = 3025$$
$$85^2 = 7225$$
$$115^2 = 13225 \quad (11 \times 12 = 132)$$
$$205^2 = 42025 \quad (20 \times 21 = 420)$$

Thus we see that the technique holds true in all the examples.

The technique of squaring numbers ending with 5 is a very popular technique. Some educational boards have included it in their curriculum. In Vedic Mathematics, there is an extension to this principle which is not known to many people. This formula of Vedic Mathematics tells us that the above rule is applicable not only to the squaring of numbers ending in 5 but also to the multiplication of numbers whose last digits add to 10 and the remaining digits are the same.

Thus, there are two conditions necessary for this multiplication. The first condition is that the last digits should add to 10 and the second condition is that the remaining digits should be the same.

Let us have a look at a few examples:

$$
\begin{array}{cccc}
6\,6 & 1\,0\,7 & 9\,1 & 5\,1 \\
\times 6\,4 & \times 1\,0\,3 & \times 9\,9 & \times 5\,9
\end{array}
$$

In the above examples, it can be observed that the last digits in each case add up to 10 and the remaining digits are the same. Let us take the first example...

Here the last digits are 6 and 4 which add up to 10. Secondly, the remaining digits are the same, viz. '6' and '6'. Thus, we can find the square of this number by the same principle which we used in squaring numbers ending with a 5.

- First, multiply the number 6 by the number that follows it. After 6 comes 7. Thus, (6×7) is 42.
- Next, we multiply the right-hand most digits (6×4) and write the answer as 24. The complete answer is 4224.

$$
\begin{array}{r}
1\,0\,7 \\
\times 1\,0\,3 \\
\hline
110\ \ 21
\end{array}
$$

In the second example, we have to multiply 107 by 103. In

this case the last digits 7 and 3 add to 10 and the remaining digits are the same. We will obtain the product using the same procedure.

- First, we will multiply the number 10 by the number that follows it, 11, and write the answer as 110
- Next, we multiply the right-hand most digits, viz. 7 and 3, and write the answer as 21. The complete answer is 11021.

In the third example, we have to multiply 91 by 99

$$
\begin{array}{r}
9\,1 \\
\times\ 9\,9 \\
\hline
90\ 09
\end{array}
$$

- We multiply 9 by the number that follows it, 10, and write the answer as 90.
- We multiply the numbers (1×9) and write the answer as 09. The final answer is 9009.

(Note: The right hand part should always be filled-in with a two-digit number. Thus, we have to convert the number 9 to 09).

In the last example, we have to multiply 51 by 59

$$
\begin{array}{r}
5\,1 \\
\times\ 5\,9 \\
\hline
30\ 09
\end{array}
$$

- We multiply 5 with the next number 6 and write the answer as 30.
- Next, we multiply (1×9) and write the answer as 09. The final answer is 3009

This formula of Vedic Mathematics works for any such numbers whose last digits add up to ten and the remaining digits are the same. The same formula works while squaring numbers ending with 5 because when you square two numbers ending with 5, then the right hand most digits add to 10 (5 plus 5) and the remaining digits are the same (since we are squaring them).

Let us look at a few other examples where the right-hand most digits add to 10 and the remaining digits are the same.

$$72 \times 78 = 5616$$

$$84 \times 86 = 7224$$

$$23 \times 27 = 621$$

$$89 \times 81 = 7209$$

$$106 \times 104 = 11024$$

$$1003 \times 1007 = 1010021$$

This was the first specific technique that we studied. The next technique that we will discuss is also related to squaring. It is used to square numbers that lie between 50 and 60.

(b) Squaring of numbers between 50 and 60

5 7	5 6	5 2	5 3
× 5 7	× 5 6	× 5 2	× 5 3

We have taken four different examples above. We will be squaring the numbers 57, 56, 52 and 53 respectively. We can find the answer to the questions by taking two simple steps as given below:

(1) Add 25 to the digit in the units place and put it as the left-hand part of the answer.

(2) Square the digits in the units place and put it as the right-hand part of the answer. (If it is a single digit then convert it to two digits)

(Q) Find the square of 57.

$$\begin{array}{r} 5\ 7 \\ \times\ 5\ 7 \\ \hline 32\ 49 \end{array}$$

- In the first example we have to square 57. In this case we add 25 to the digit in the units place, viz., 7. The answer is 32 which is the LHS (left-hand side) of our answer. (Answer at this stage is 32___)
- Next, we square the digit in the units place '7' and get the answer as 49. This 49 we put as the right hand part of our answer. The complete answer is 3249.

(Q) Find the square of 56.

$$\begin{array}{r} 5\ 6 \\ \times\ 5\ 6 \\ \hline 31\ \ 36 \end{array}$$

In the second example, we add 25 to 6 and get the LHS as 31. Next, we square 6 and get the answer 36 which we put on the RHS. The complete answer is 3136.

(Q) Find the square of 52.

$$\begin{array}{r} 5\ 2 \\ \times\ 5\ 2 \\ \hline 27\ \ 04 \end{array}$$

In the third example, we add 2 to 25 and get the LHS as 27. Next, we square 2 and get the answer 4 which we will put on the RHS. However, the RHS should be a two-digit number. Hence, we convert 4 to a two-digit number and represent it as 04. The complete answer is 2704.

(Q) Find the square of 53.

$$\begin{array}{r} 5\ 3 \\ \times\ 5\ 3 \\ \hline 28\ \ 09 \end{array}$$

In the last example, we add 3 to 25 and get the answer as 28. Next, we square 3 and get the answer as 9. As mentioned in rule B, the answer on the RHS should be converted to two digits. Thus, we represent the digit 9 as 09. The complete answer is 2809.

On similar lines we have:

$$51^2 = 2601$$
$$52^2 = 2704$$
$$54^2 = 2916$$
$$55^2 = 3025$$
$$58^2 = 3364$$

(c) Multiplication of numbers with a series of 9's

In my seminars, I often have an audience challenge round. In this round, the audience members ask me to perform various mental calculations and give them the correct answer. They generally ask me to multiply numbers which involve a lot of 9's in them. The general perception is that the higher the number of 9's the tougher it will be for me to calculate. However, the truth is exactly the opposite — the higher the number of 9's in the question, the easier it is for me to calculate the correct answer. I use two methods for this. The first method is given below and the second method is explained in the chapter 'Base Method of Multiplication'.

Using the method given below, we can multiply any given number with a series of nines. In other words, we can instantly multiply any number with 99, 999, 9999, 99999, etc.

The technique is divided into three cases. In the first case, we will be multiplying a given number with an equal number of nines. In the second case we will be multiplying a number with a higher number of nines. In the third case, we will be multiplying a number with a lower number of nines.

Case 1

(Multiplying a number with an equal number of nines)

(Q) Multiply 654 by 999.

$$\begin{array}{r} 654 \\ \times\ 999 \\ \hline 653\ 346 \end{array}$$

- We subtract 1 from 654 and write half the answer as 653. Answer at this stage is 653____
- Now we will be dealing with 653. Subtract each of the digits six, five and three from nine and write them in the answer one by one.

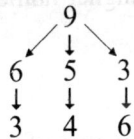

- Nine minus six is 3. Nine minus five is 4. Nine minus three is 6.
- The answer already obtained was 653 and now we suffix to it the digits 3, 4 and 6. The complete answer is 653346.

(Q) Multiply 9994 by 9999.

$$\begin{array}{r} 9\ 9\ 9\ 4 \\ \times\ 9\ 9\ 9\ 9 \\ \hline 9993\ 0006 \end{array}$$

We subtract one from 9994 and write it as 9993. This becomes our left half of the answer. Next, we subtract each of the digits of 9993 from 9 and write the answer as 0006. This becomes the right half of the answer. The complete answer is 99930006.

(Q) Multiply 456789 by 999999.

$$\begin{array}{r} 456789 \\ \times\ 999999 \\ \hline 456788\ 543211 \end{array}$$

We subtract 1 from 456789 and get the answer 456788. We write this down on the left hand side. Next, we subtract each of the digits of 456788 (left hand side) from 9 and get 543211 which becomes the right hand part of our answer. The complete answer is 456788543211.

More examples:

7777	65432	447	90909
× 9999	× 99999	× 999	× 99999
7776 2223	65431 34568	446 553	90908 09091

The simplicity of this method can be vouched from the examples given above. Now we move to Case 2. In this case, we will multiply a given number with a higher number of nines.

Case 2

(Multiplying a number with a higher number of nines)

(Q) Multiply 45 with 999.

$$\begin{array}{r} 45 \\ \times\ 999 \end{array} \longrightarrow \begin{array}{r} 045 \\ \times\ 999 \\ \hline 044955 \end{array}$$

There are three nines in the multiplier. However, the multiplicand 45 has only two digits. So we add a zero and convert 45 to 045 and make it a three digit number. After having done so, we can carry on with the procedure explained in Case 1.

First we subtract 1 from 045 and write it down as 044. Next, we subtract each of the digits of 044 from 9 and write the answer as 955. The complete answer is 044955.

(Q) Multiply 888 with 9999.

$$\begin{array}{r} 888 \\ 9999 \end{array} \longrightarrow \begin{array}{r} 0888 \\ 9999 \\ \hline 8879112 \end{array}$$

We convert 888 to 0888 and make the digits equal to the number of nines in the multiplier. Next, we subtract 1 from 0888 and write the answer as 0887. Finally, we subtract each digit of 0887 from 9 and write the answer as 9112. The final answer is 08879112 which is 8879112.

(Q) Multiply 123 by 99999.

$$\begin{array}{r} 123 \quad\longrightarrow\quad 00123 \\ \times\ 99999 \qquad\qquad \times\ 99999 \\ \hline 00122\ /\ 99877 \end{array}$$

The multiplicand is a three-digit number and the multiplier is a five-digit number. Therefore, we add two zeros in the multiplicand so that the digits are equal in the multiplicand and the multiplier.

We now subtract 1 from 00123 and write the left hand part of the answer as 00122. Next, we subtract each of the digits of the left hand part of the answer from 9 and write it down as 99877 as the right hand part of the answer. The complete answer is 12299877.

Other examples:

162	5555	363	10101
× 9999	× 99999	× 999999	× 9999999
0161 / 9838	05554 / 94445	000362 / 999637	0010100 / 9989899

We can see that this technique is not only simple and easy to follow, but it also enables one to calculate the answer in the mind itself. This is the uniqueness of these systems. As you read the chapters of this book, you will realize how simple and easy it is to find the answer to virtually any problem of mathematics that one encounters in daily life and especially in the exams. And the approach is so different from the traditional methods of calculation that it makes the whole process enjoyable.

Case 3 of this technique deals with multiplying a number with a lower number of nines. There is a separate technique for this in Vedic Mathematics and requires the knowledge of the Nikhilam Sutra (explained later in this book). However, at this point of time, we can solve such problems using our normal practices of instant multiplication.

(Q) Multiply 654 by 99.

In this case the number of digits are more than the number of nines in the multiplier. Instead of multiplying the number 654 with 99 we will multiply it with (100-1). First we will multiply 654 with 100 and then we will subtract from it 654 multiplied by 1.

$$654 \times 99$$

$$65400$$
$$-654$$
$$= 64746$$

(Q) Multiply 80020 by 999.

We will multiply 80020 with (1000 - 1).

$$80020000$$
$$-80020$$
$$= 79939980$$

This method is so obvious that it needs no further elaboration.

(d) Multiplication of numbers with a series of 1's

In the previous technique we saw how to multiply numbers with a series of 9's. In this technique we will see how to multiply numbers with a series of 1's. Thus, the multiplier will have numbers like 1, 11, 111, etc.

Let us begin with the multiplier 11.

(Q) Multiply 32 by 11.

$$\begin{array}{r} 3\ 2 \\ \times\ 1\ 1 \\ \hline 352 \end{array}$$

- First we write the right-hand most digit 2 as it is. (Answer = _____2)
- Next, we add 2 to the number in left 3 and write 5. (Answer = _____52)
- Last, we write the left hand most digit 3 as it is.

Thus, the answer is 352.

(Q) Multiply 43 by 11.

$$\begin{array}{r} 4\ 3 \\ \times\ 1\ 1 \\ \hline 473 \end{array}$$

Write the last digit 3 as it is. Next we add 3 to 4 and get 7. Finally we write 4 as it is. The complete answer is 473.

(Q) Multiply 64 by 11.

$$\begin{array}{r} 6\ 4 \\ \times\ 1\ 1 \\ \hline 704 \end{array}$$

In this example we write down the last digit 4 as it is. Next, we add 4 to 6 and get the answer 10. Since, 10 is a two-digit answer, we write down the 0 and carry over 1. Finally, we add 1 to 6 and make it 7. The complete answer is 704.

(Q) Multiply 652 by 11.

$$\begin{array}{r} 6\ 5\ 2 \\ \times\ 1\ 1 \\ \hline 7172 \end{array}$$

The logic of two digit numbers can be expanded to higher numbers. In the given example we have to multiply 652 by 11.

- We write down the last digit of the answer as 2. (Answer = ____2)
- Next, we add 2 to 5 and make it 7. (Answer = ____72)
- Next, we add 5 to 6 and make it 11. We write down 1 and carry over 1. (Answer is ____172)
- Last, we take 6 and add the one carried over to make it 7. (Final answer is 7172).

(Q) Multiply 3102 by 11.

$$
\begin{array}{r}
3\ 1\ 0\ 2 \\
\times\ 1\ 1 \\
\hline
34122
\end{array}
$$

- We write down 2 as it is. (Answer is ____2)
- We add 2 to 0 and make it 2 (Answer is ____22)
- We add 0 to 1 and make it 1. (Answer is ____122)
- We add 1 to 3 and make it 4 (Answer is ____4122)
- We write the first digit 3 as it is (Final answer is 34122)

Similarly,

$$
\begin{array}{cccc}
4\ 1 & 303 & 1309 & 2901265 \\
\times\ 1\ 1 & \times\ 11 & \times\ 11 & \times\ \ \ \ \ 11 \\
\hline
451 & 3333 & 14399 & 31913915
\end{array}
$$

When we multiply a number by 11 we write the last digit as it is. Then we move towards the left and continue to add two digits at a time till we reach the last digit which is written as it is.

Since the multiplier 11 has two 1's we add maximum two digits at a time. When the multiplier is 111 we will add maximum three digits at a time because the multiplier 111 has three digits. When the multiplier is 1111 we will add maximum four digits at a time since the multiplier 1111 has four digits.

We have already seen how to multiply numbers by 11. Let us have a look at how to multiply numbers by 111.

(Q) Multiply 203 by 111.

$$\begin{array}{r} 2\ 0\ 3 \\ \times\ 1\ 1\ 1 \\ \hline 22533 \end{array}$$

- We write down the digit in the units place 3 as it is in the answer
- We move to the left and add $(3 + 0) = 3$
- We move to the left and add $(2 + 0 + 3) = 5$ (maximum three digits)
- We move to the left and add $(0 + 2) = 2$
- We take the last digit 2 and write down as it is

(Q) Multiply 201432 by 111.

$$\begin{array}{r} 201432 \\ \times\ \ \ 111 \\ \hline 22358952 \end{array}$$

- The (2) in the units place of the multiplicand is written as the units digit of the answer
- We move to the left and add $(2 + 3)$ and write 5
- We move to the left and add $(2 + 3 + 4)$ and write 9
- We move to the left and add $(3 + 4 + 1)$ and write 8
- We move to the left and add $(4 + 1 + 0)$ and write 5
- We move to the left and add $(1 + 0 + 2)$ and write 3
- We move to the left and add $(0 + 2)$ and write 2
- We move to the left and write the single digit (2) in the answer.

Thus, the complete answer obtained by each of the steps above is 22358952.

Similarly,

111	2035	90321	6021203
× 111	× 111	× 111	× 111
12321	225885	10025631	668353533

The simplicity of this method is evident from the examples. In most cases you will get the answer within a minute. In fact, the beauty of this technique is that it converts a process of multiplication to basic addition.

Using the same method, we can multiply any number by a series of 1's.

If you want to multiply a number by 1111 you can use the method given above. The only difference will be that we will add maximum four numbers at a time (because there are four ones in 1111) and when the multiplier is 11111 we will be multiplying maximum five digits at a time. An example of the former type is given below:

(Q) Multiply 210432 by 1111.

$$\begin{array}{r} 210432 \\ \times\ 1111 \\ \hline 233789952 \end{array}$$

- We write down the last digit 2 as it is — 2
- Add (2 + 3) = 5 — 52
- Add (2 + 3 + 4) = 9 — 952
- Add (2 + 3 + 4 + 0) = 9 — 9952
- Add (3 + 4 + 0 + 1) = 8 — 89952
- Add (4 + 0 + 1 + 2) = 7 — 789952
- Add (0 + 1 + 2) = 3 — 3789952
- Add (1 + 2) = 3 — 33789952
- Add (2) = 2 — 233789952

Thus, the product of (210432 × 1111) = 233789952.

(e) Multiplication of numbers with a series of similar digits in multiplier

This technique is basically an extension of the previous technique. In technique 'c' we saw how to multiply a number with a series of nines and in technique 'd' we saw how to multiply a number with a series of ones.

A question may arise regarding how to multiply numbers with a series of 2's, like 2222, or with a series of 3's, like 333, and such other numbers.

Let us have a look at a few examples:

(Q) Multiply 333 by 222.

The question asks us to multiply 333 by 222. Now, carefully observe the logic that we apply in this case

$$333 \times 222$$
$$= \quad 333 \times 2 \times 111 \text{ (because 222 is 2 multiplied by 111)}$$
$$= \quad 666 \times 111 \text{ (because 333 multiplied by 2 is 666)}$$

Therefore, multiplication of 333 by 222 is the same as multiplication of 666 by 111. But, we have already studied the procedure of multiplying a number by 111 in the previous sub-topic. Our answer will be as follows:

$$
\begin{array}{r}
6\,6\,6 \\
\times\ 1\,1\,1 \\
\hline
7\,3\,9\,2\,6 \\
\end{array}
$$

Therefore, 333×222 is 73926

(Q) Multiply 3021 by 333.

The multiplicand is a normal number and the multiplier is a series of 3's. We do not know how to multiply a number with a series of 3's but we know how to multiply a number with a series of 1's. Thus, we will represent the expression in such a manner that the multiplier is 111.

3021×333

$= 3021 \times 3 \times 111$ (because 333 is same as 3 into 111)

$= 9063 \times 111$ (because 3021 into 3 is 9063.

We have already learnt how to multiply 9063 by 111. On the basis of our knowledge we can easily complete the multiplication

$$
\begin{array}{r}
9\ 0\ 6\ 3 \\
\times\ 1\ 1\ 1 \\
\hline
1005993
\end{array}
$$

On the basis of the above examples it can be seen that there is no need of explaining the procedure of multiplication. The procedure is the same as observed in the previous sub-topic. Basically, we have to convert a series of 2's, 3's, 4's, etc. in the multiplier to a series of 1's by dividing it by a certain number. Next, we have to multiply the multiplicand by the same number

(a) Multiply 1202 by 44.

$$
\begin{array}{rcl}
1202 & \longrightarrow & 4808 \\
\times\ 44 & & \times\ 11 \\
& & \hline \\
& & 52888
\end{array}
$$

In this case, we have divided the multiplier 44 by 4 to obtain a series of 1's (11). Since we have divided the multiplier by 4 we will multiply the multiplicand 1202 by 4. Thus, we have the new multiplicand as 4808. When 4808 is multiplied by 11 the answer is 52888. This is also the answer to the original question of 1202 by 44.

(b) Multiply 2008 by 5555.

$$
\begin{array}{rcl}
2008 & \longrightarrow & 10040 \\
5555 & & \times\ 1111 \\
& & \hline \\
& & 11154440
\end{array}
$$

Ans: The product of 2008 multiplied by 5555 is 11154440

(c) Multiply 10503 by 888.

$$10503 \longrightarrow 84024$$
$$\underline{\times\ 888} \qquad\qquad \underline{\times\ 111}$$
$$9326664$$

Ans: The product of 10503 multiplied by 888 is 9326664.

(f) Subtraction using the rule 'All from 9 and the last from 10'

Subtraction using the rule 'All from 9 and the last from 10' is one of the elementary techniques of Vedic Mathematics. Basically, it is used to subtract any number from a power of ten. The powers of ten include numbers like 10, 100, 1000, 10000, etc.

So if you want to learn a method by which you can quickly subtract a number from a power of ten, then this technique can come to your aid.

When we go to the market to buy something, we generally give a hundred rupee note to the shopkeeper and calculate the change that we should get after deducting the total amount of groceries. In such a situation, this technique can come to our aid.

(Q) Subtract 54.36 from 100.

$$1\ 0\ 0.0\ 0$$
$$-\ \underline{5\ 4.3\ 6}$$

We are asked to subtract 54.36 from 100. In this case, we generally start from the right and subtract 6 from 0. But, we realize that it is not possible to subtract 6 from 0 and so we move to the number in the left and then borrow one and give it to zero and make it ten and so on.

This whole procedure is slightly cumbersome and there is a possibility of making a mistake too.

Vedic Mathematics provides a very simple alternative. The approach of Vedic Mathematics is explained by the rule 'All from 9 and the last from 10.' It means that we have to subtract each digit from nine and subtract the last digit from 10. This will give us the answer.

The number to be subtracted is 54.36. We have to subtract all the digits from nine except for the last digit which will be subtracted from ten. Thus,

$$9 - 5 = 4$$
$$9 - 4 = 5$$
$$9 - 3 = 6$$
$$10 - 6 = 4$$

The final answer is 45.64.

(Q) Subtract 3478.2281 from 10000.

$$\begin{array}{r} 10000.0000 \\ -\ 3478.2281 \end{array}$$

In this case, we will subtract the digits 3, 4, 7, 8, 2, 2 and 8 from 9 and the last digit 1 from 10. The respective answers will be 6, 5, 2, 1, 7, 7, 1 and 9. Thus, the final answer is 6521.7719

More examples:

(a) $\begin{array}{r} 1000.000 \\ -\ 363.633 \\ \hline 636.367 \end{array}$ (b) $\begin{array}{r} 10000.00 \\ -\ 9191.09 \\ \hline 808.91 \end{array}$ (c) $\begin{array}{r} 10.000000 \\ -\ 7.142857 \\ \hline 2.\ 857143 \end{array}$

(d) $\begin{array}{r} 10000 \\ -\ 23 \end{array} \longrightarrow \begin{array}{r} 10000 \\ -\ 0023 \\ \hline 9977 \end{array}$

(e) $\begin{array}{r} 100000.00 \\ -\ 459.62 \end{array} \longrightarrow \begin{array}{r} 100000.00 \\ -\ 00459.62 \\ \hline 99540.38 \end{array}$ (f) $\begin{array}{r} 100.00 \\ -\ 17.10 \\ \hline 82.90 \end{array}$

The examples prove the simplicity and efficiency of this system. In examples (d) and (e) we added zeros in the number below so that we get the accurate answer.

In this chapter we have seen 6 simple yet quick techniques of Vedic Mathematics. I wanted to begin the book with these easy techniques so that we can prepare ourselves for the comprehensive techniques that will follow in the forthcoming chapters.

You must have observed that the techniques that we have employed in this chapter work with a totally different approach. We have found answers to our questions using a completely different approach. In all the chapters of this book you will discover that the method used in solving the problems is far more efficient than the normal systems that we have been using, thus enabling us to produce outstanding results.

EXERCISE

Q. (1) Find the product in the following numbers whose last digits add to ten.

  (a) 45 × 45
  (b) 95 × 95
  (c) 111 × 119
  (d) 107 × 103

Q. (2) Find the squares of the following numbers between fifty and sixty.

  (a) 56
  (b) 51
  (c) 53

Q. (3) Find the product of the following numbers which are multiplied by a series of nines.

  (a) 567 × 999
  (b) 23249 × 99999
  (c) 66 × 9999
  (d) 302 × 99999

Q. (4)   Find the product of the following numbers which are multiplied by a series of ones.

   (a) 32221 × 11
   (b) 64609 × 11
   (c) 12021 × 111
   (d) 80041 × 111

Q. (5)   Find the product of the following numbers which are multiplied by a series of same numbers.

   (a) 7005 × 77
   (b) 1234 × 22
   (c) 2222 × 222
   (d) 1203 × 333

Q. (6)   Subtract the following numbers from a given power of ten.

   (a) 1000 minus 675.43
   (b) 10000 minus 7609.98
   (c) 10000 minus 666
   (d) 1000 minus 2.653

CHAPTER 2

Criss-Cross System of Multiplication

The traditional system of multiplication taught to students in schools and colleges is a universal system, i.e., it is applicable to all types of numbers. The traditional system can also be used for numbers of any length.

Let us have a look at an example:

```
          2   3   4
      ×   3   6   2
      ─────────────
          4   6   8
      1   4   0   4   0
  +   7   0   2   0   0
  ─────────────────────
      8   4   7   0   8
```

This is the traditional way of multiplication which is taught to students in schools. This system of multiplication is perfect and works for any combination of numbers.

In Vedic Mathematics too, we have a similar system but it helps us to get the answer much faster. This system is also a universal system and can be used for any combination of numbers of any length.

This system of multiplication is given by the 'Urdhva-Tiryak Sutra.' It means 'vertically and cross-wise'. The applications of this system are manifold, but in this chapter we shall confine our study only to its utility in multiplying numbers. We shall call it the *Criss-Cross system of multiplication*.

METHOD

Let us suppose we want to multiply 23 by 12. According to the traditional system, our answer would have been:

$$
\begin{array}{r}
2 \quad 3 \\
\times \; 1 \quad 2 \\
\hline
4 \quad 6 \\
+\,2 \quad 3 \quad 0 \\
\hline
2 \quad 7 \quad 6 \\
\end{array}
$$

With the Criss-Cross system, we can get the answer in just one step as given below:

$$
\begin{array}{r}
2 \quad 3 \\
\times \; 1 \quad 2 \\
\hline
2 \quad 7 \quad 6 \\
\end{array}
$$

Let us have a look at the modus operandi of this system:

Step 1   2   3
                        ↓
              × 1   2
              _____
                        6

We multiply the digits in the ones place, that is, $3 \times 2 = 6$. We write 6 in the ones place of the answer.

Step 2

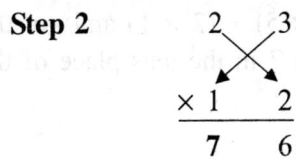

Now, we cross multiply and add the products, that is, $(2 \times 2) + (3 \times 1) = 7$. We write the 7 in the tens place of the answer.

Step 3

```
     2   3
      ↓
 ×   1   2
 2   7   6
```

Now we multiply the ones digits, that is, $2 \times 1 = 2$.

The completed multiplication is:

```
     2   3
 ×   1   2
 2   7   6
```

Let us notate the three steps involved in multiplying a 2-digit number by a 2-digit number.

(a) ```
    *
    *
    ``` (b) ```
    *   *
    *   *
    ``` (c) ```
    *
    *
    ```

We shall have a look at one more example.

Let us multiply 31 by 25

```
    3   1
 ×  2   5
        5
```

- First we multiply 1 by 5 vertically and get the answer as 5

```
    3   1
 ×  2   5
    7   5
```
```

- Then, we cross-multiply $(3 \times 5) + (2 \times 1)$ and get the answer as 17. We write down 7 in the tens place of the answer and carry over 1.

$$
\begin{array}{rr}
3 & 1 \\
\times\ 2 & 5 \\
\hline
\mathbf{7} \qquad 7 & 5
\end{array}
$$

- Lastly, we multiply (3×2) and get the answer as 6. But, we have carried over 1. So, the final answer is 7.

Given below are a few examples of 2-digit multiplication where there is no carrying-over involved:

(A)	(B)	(C)
1 2	2 1	3 2
2 1	1 4	1 1
2 : 1+4 : 2	2 : 1+8 : 4	3 : 2+3 : 2
= **252**	= **294**	= **352**

An example of 2-digit numbers where there is a carry-over involved:

$$
\begin{array}{rr}
2 & 3 \\
\times\ 1 & 4 \\
\hline
3 \quad 2 & 2
\end{array}
$$

- First, we multiply 3 by 4. The answer is 12. We write down 2 and carry over 1. The answer at this stage is _____2.
- Next, we cross multiply (2×4) and add it to (3×1). The total is 11. Now, we add the 1 which we carried over. The total is 12. So, we write 2 and carry over 1.
  (The answer at this stage is _____22).
- Last, we multiply (2×1) and get the answer as 2. To it, we add the 1 that is carried over and get the final answer as 3.

(The completed answer is 322).

Thus, we see how the Criss-Cross system of multiplication helps us get our answer in just one line! And, the astonishing fact is that this same system can be expanded to multiplication of numbers of higher digits too.

And in every case, we will be able to get the answer in a single line.

Let us have a look at the multiplication process involved in multiplying a three-digit number by another three-digit number.

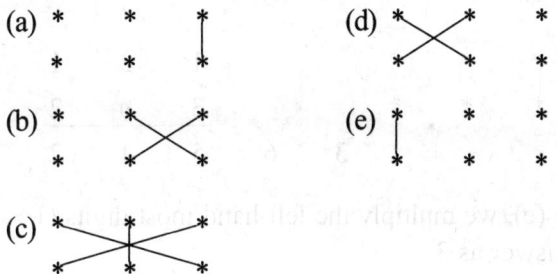

Let us multiply two three-digit numbers where there is no carry over involved.

(a)
```
*   *   |
*   *   *
```
1   2   1
3   0   2
_____
        2

As suggested by step (a), we multiply 1 into 2 and get the answer as 2.

(b)
```
*   ✕
*   ✕
```
1   2   1
3   0   2
_____
    4   2

Next, we cross-multiply (2 × 2) and add it to (1 × 0). Thus, the final answer is 4.

(c)
```
*✕*✕*
*✕*✕*
```
1   2   1
3   0   2
_____
5   4   2

In step (c), we multiply (1×2) and (2×0) and (3×1). We add the three answers thus obtained to get the final answer 5.

(d)

			1	2	1
			3	0	2
		6	5	4	2

In step (d), we multiply (1×0) and (3×2). The final answer is 6.

(e)

			1	2	1
			3	0	2
	3	6	5	4	2

In step (e), we multiply the left-hand most digits (1×3) and get the answer as 3.

Thus, it can be seen that the product obtained by multiplying two 3-digit numbers can be obtained in just one line with the help of the Criss-Cross system.

We shall quickly have a look at how to multiply two 3-digit numbers where there is a carry over involved. Obviously, the process of carrying over is the same as we use in normal multiplication.

Example:

	1	2	4
× 3	5	5	

	4	4	0	2	0

- We multiply 4 by 5 and get the answer as 20. We write down 0 and carry over 2.
  (The answer at this stage is _____ 0)

- (2×5) is 10 plus (4×5) is 20. The total is 30 and we add the 2 carried over to get 32. We write down 2 and carry over 3.
  (The answer at this stage is _____ 20)

- (1 × 5) is 5 plus (2 × 5) is 10 plus (4 × 3) is 12. The total is 27 and we add the 3 carried over to get the answer as 30. We write down 0 and carry over 3.
  (The answer at this stage is ____020)

- (3 × 2) is 6 plus (1 × 5) is 5. The total is 11 and we add the 3 carried over. The final answer is 14. We write down 4 and carry over 1.
  (The answer at this stage is _____4020)

- Finally, (1 × 3) is 3. Three plus 1 carried over is 4. The final answer is 44020.

A few examples with their completed answers are given below:

3 4 2	4 1 2	5 5 5	3 9 1
×5 0 6	×9 0 3	×2 2 2	×2 7 4
173052	372036	123210	107134

MULTIPLICATION OF HIGHER-ORDER NUMBERS

We have seen how to multiply 2-digit and 3-digit numbers. We can expand the same logic and multiply bigger numbers. Let us have a look at how to multiply 4-digit numbers.

The following are the steps involved in multiplying 4-digit numbers:

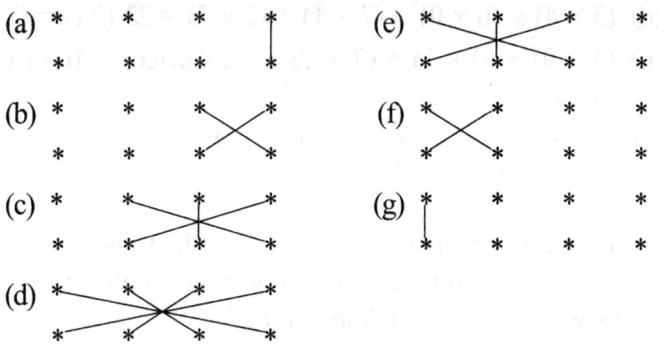

Suppose we want to multiply 1111 by 1111. Then, there will be 7 steps involved in the complete multiplication as suggested above from steps a to g.

(a)	(b)	(c)	(d)
1 1 1 **1**	1 1 **1 1**	1 **1 1 1**	**1 1 1 1**
× 1 1 1 **1**	× 1 1 **1 1**	× 1 **1 1 1**	× **1 1 1 1**
1	**2** 1	**3** 2 1	**4** 3 2 1

(e)	(f)	(g)
1 1 1 1	1 **1** 1 1	1 1 **1** 1
× **1** 1 1 1	× 1 **1** 1 1	× 1 1 **1** 1
3 4 3 2 1	**2** 3 4 3 2 1	**1** 2 3 4 3 2 1

Let us have a look at one more example:

Example: 2104 multiplied by 3072.

$$
\begin{array}{r}
2\ 1\ 0\ 4 \\
\times\ 3\ 0\ 7\ 2 \\
\hline
\mathbf{6\ 4\ 6\ 3\ 4\ 8\ 8}
\end{array}
$$

STEP-WISE ANSWERS

(a) $(4 \times 2) = \mathbf{8}$

(b) $(7 \times 4) + (0 \times 2) = \mathbf{28}$ (2 carry-over)

(c) $(0 \times 4) + (0 \times 7) + (1 \times 2) + (2 \text{ carried}) = \mathbf{4}$

(d) $(3 \times 4) + (0 \times 0) + (7 \times 1) + (2 \times 2) = \mathbf{23}$ (2 carry over)

(e) $(3 \times 0) + (0 \times 1) + (7 \times 2) + (2 \text{ carried}) = \mathbf{16}$ (1 carry over)

(f) $(2 \times 0) + (3 \times 1) + (1 \text{ carried}) = \mathbf{4}$

(g) $(2 \times 3) = \mathbf{6}$

We have seen the multiplication of 2-digit, 3-digit and 4-digit numbers. A question may arise regarding multiplication of numbers with an unequal number of digits.

Let us suppose you want to multiply 342 by 2009. Here, we have one number that has three digits and another number that has four digits. Now, for such a multiplication which technique will you use?

Will you use the technique used for multiplying numbers of 3-digits or the technique used for multiplying numbers of 4 digits?

To multiply 342 by 2009, write the number 342 as 0342 and then multiply it by 2009. Use the technique used for multiplying four-digit numbers.

$$\begin{array}{r} 3\ 4\ 2 \\ \times\ 2\ 0\ 0\ 9 \\ \hline \end{array} \longrightarrow \begin{array}{r} 0\ 3\ 4\ 2 \\ \times\ 2\ 0\ 0\ 9 \\ \hline \end{array}$$

Thus, if we want to multiply 312 by 64, we will write 64 as 064 and then multiply it by 312 using the technique of 3-digit multiplication.

THE CHARACTERISTICS OF CRISS-CROSS MULTIPLICATION

Now, we shall learn a few characteristics of the Criss-Cross (urdhva-tiryak) system of multiplication. The knowledge of these characteristics will help us to easily calculate any answer.

To understand the characteristics, we shall carefully observe the steps used in multiplying 2-digit, 3-digit and 4-digit numbers. Let us have a look at them:

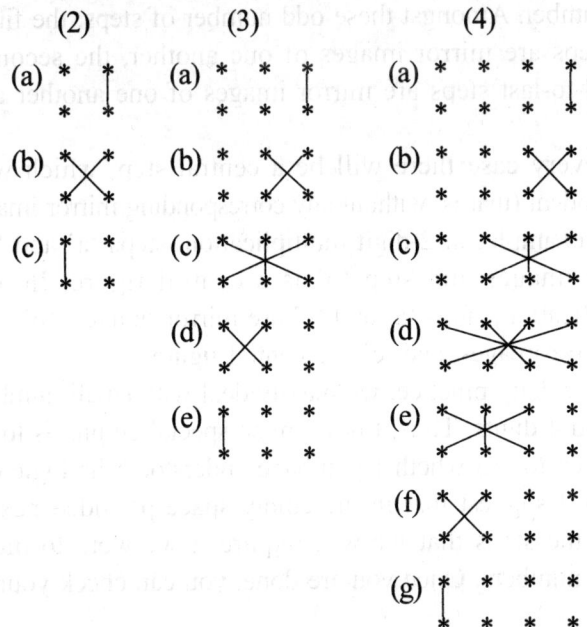

The steps used for multiplying 2-digit, 3-digit and 4-digit numbers is 3, 5, and 7 respectively. The number of steps used for any multiplication can be found using the formula:

    '2 multiplied by (number of digits) minus 1'

Thus, when we multiply 2-digit numbers, the steps used are $2 \times 2 - 1 = 3$ and therefore we use 3 steps. When we multiply 4-digit numbers, the steps used are $2 \times 4 - 1 = 7$. When we multiply 5-digit numbers, the steps used will be $2 \times 5 - 1 = 9$.

(If there are an unequal number of digits in the multiplicand and the multiplier, they should first be made equal by inserting 0's at the appropriate places and then the formula must be used.)

Most students will be able to guess the trend of the steps used in this system of multiplication by the mere observation of the examples that we have solved. This is because they follow a systematic pattern.

The second characteristic of the Criss-Cross system of multiplication is that the number of steps used will always be an odd number. Amongst these odd number of steps, the first and last steps are mirror images of one another, the second and second-to-last steps are mirror images of one another and so on.

In every case there will be a central step, which will be independent (that is, without any corresponding mirror image.)

For example, in 2-digit multiplication, steps 'a' and 'c' are mirror images and step 'b' is a central figure. In 3-digit multiplication, steps 'a' and 'e' are mirror images, 'b' and 'd' are mirror images, and 'c' is a central figure.

In our daily practice, we mainly deal with small numbers of 2, 3 and 4-digits. Thus, I have given special emphasis to them. However, to test whether you have understood the logic of this system, I suggest that in the empty space provided next, you notate the steps that we will require if we were to multiply 5-digit numbers. Once you are done, you can check your work with the answer mentioned in APPENDIX A at the end of the book.

(a) * * * * *
     * * * * *

(b) * * * * *
     * * * * *

(c) * * * * *
     * * * * *

(d) * * * * *
     * * * * *

(e) * * * * *
     * * * * *

(f) * * * * *
     * * * * *

(g) * * * * *
     * * * * *

(h) * * * * *
     * * * * *

(i) * * * * *
     * * * * *

Don't proceed until you have checked the answer in APPENDIX A.

Most of you might have made a mistake in step 'e' and solved the other steps correctly. If you got all the steps correct then it is indeed praiseworthy. Those who made a mistake in other steps need not get disheartened: with enough practice you shall be able to solve sums effortlessly.

I would again like to draw your attention to the trend of the steps used in the multiplication procedure. Since we are multiplying two five-digit numbers, the number of steps according to the formulae — 2 multiplied by the number of digits minus 1 — will be 9, and thus we have used steps from (a) to (i).

Further, you will observe that steps (a) and (i) are mirror-images, steps (b) and (h) are mirror-images, steps (c) and (g) are mirror images, steps (d) and (f) are mirror images and step (e) is a central figure.

The Criss-Cross system not only helps us to get answers quickly but also helps us to eliminate all the intermediate steps used in the multiplication process. This quality of the system can be of immense aid to students giving various competitive and professional exams.

Let us take a hypothetical situation:

We know that in competitive exams, we are given a question with four alternatives out of which we have to select the correct one.

(Q) What is the product of 121 by 302?

(a) 36522   (b) 36592   (c) 36542   (d) 36544

In this question we are asked to calculate the product of 121 by 302. Now, I read the question and start the multiplication process using the Criss-Cross system.

$$
\begin{array}{r}
1\ \ 2\ \ 1 \\
\times\ 3\ \ 0\ \ 2 \\
\hline
\end{array}
$$

First, I multiply the extreme digits 1 by 2 and get the final answer as 2. My answer at this stage is ____2.

Next, I cross multiply (2 × 2) and add it to (1 × 0). My final answer is 4. The answer at this stage is ____42.

The moment I get the digits 4 and 2 as the last two digits of my answer, I discontinue the multiplication process and instantly tick option (c) as the correct answer to the problem.

The reason is that the last two digits of the other alternatives are 22, 92 and 44 but I know the last two digits of my answer are 42 and so the correct answer can only be option (c).

This was just an example. The idea which I am trying to convey is that while solving any multiplication problem, the moment you get a part of the answer which is unique to any one of the given alternatives, you can instantly mark that alternative as the correct answer.

This advantage is not available with the traditional system as it compels you to do the whole multiplication procedure with the intermediary steps.

Traditional Method	Vedic Method

Traditional Method:
```
        1  2  1
     ×  3  0  2
     ─────────
        2  4  2
     0  0  0  0
×  3  6  3  0  0
───────────────
   3  6  5  4  2
```

Vedic Method:
```
     1  2  1
  ×  3  0  2
  ──────────
3  6  5  4  2
```

This comparison proves the fact that the urdhva-tiryak sutra (which I call the Criss-Cross system) helps us to solve any multiplication problem instantly. There are numerous methods in mathematics which help you to multiply numbers quickly but most of them can be used only for a particular category of numbers, like numbers adding up to 10, numbers between 11 and 20, etc. However, this system is a universal system and can be used for any combination of numbers.

In the next page, I have given the steps for multiplying six-digit and seven-digit numbers. In this manner we can go on and on and on, with eight-digit, nine-digit, ten-digit numbers, etc., but I guess the techniques that we have studied up to this point will suffice us to get a thorough understanding of the concept.

The concept that we have studied here can be applied in multiplying algebraic expressions too. For example, if you want to find the product of $a + 6b$ multiplied by $2a + 3b$ you can use the Criss-Cross system. For solved examples refer to Appendix B.

TECHNIQUE FOR MULTIPLYING SIX-DIGIT NUMBERS

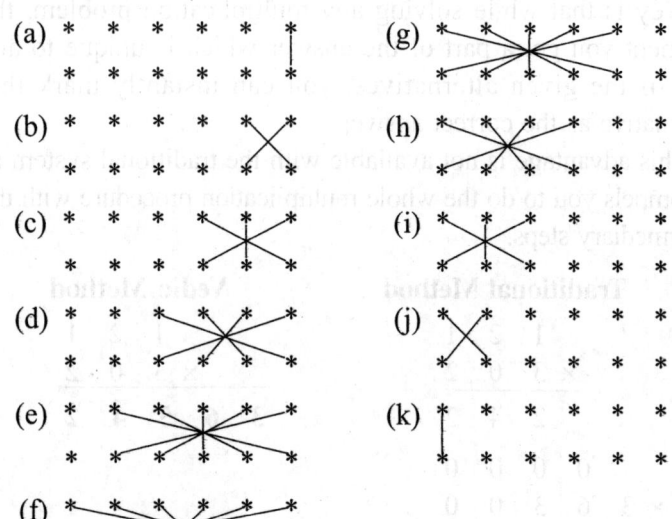

TECHNIQUE FOR MULTIPLYING SEVEN-DIGIT NUMBERS

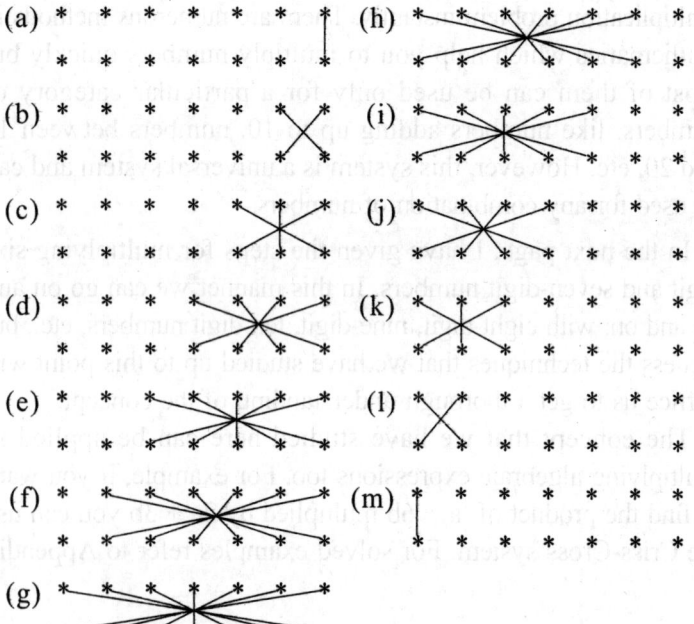

EXERCISE 1

Questions in PART A are very easy questions and you will be able to solve them just by looking at the numbers without any rough work.

Questions in PART B are of average difficulty and you will be able to solve them with some thought.

Questions in PART C will compel you to do some writing work to get the answer.

PART A

(1) 2 3	(2) 3 4	(3) 3 3	(4) 4 1
× 1 2	× 1 1	× 2 1	× 1 3

(5) 2 1 1	(6) 2 2 2	(7) 3 0 3	(8) 1 1 1 1
× 3 2 0	× 1 1 1	× 2 1 0	× 1 1 1 1

PART B

(1) 4 4	(2) 3 3	(3) 9 1	(4) 2 4
× 2 2	× 4 1	× 3 1	× 5 1

(5) 3 5 8	(6) 4 2 3	(7) 8 0 1	(8) 2 3 2 3
× 1 1 1	× 2 0 2	× 6 0 1	× 3 2 3 2

PART C

(1) 6 8	(2) 3 5	(3) 6 2	(4) 9 1
× 3 2	× 4 1	× 2 6	× 8 2

(5) 7 2 1	(6) 4 2 9	(7) 8 7 6	(8) 4 3 2 1
× 2 1 9	× 9 1 1	× 3 0 7	× 9 1 0 1

Squaring Numbers

Squaring can be defined as 'multiplying a number by itself.'

There are many different ways of squaring numbers. Many of these techniques have their roots in multiplication as squaring is simply a process of multiplication.

Examples: 3^2 is 3 multiplied by 3 which equals 9

4^2 is four multiplied by 4 which equals 16

The techniques that we will study are:

(A) Squaring of numbers using the Criss-Cross system

(B) Squaring of number using formulae

(A) SQUARING OF NUMBERS USING CRISS-CROSS SYSTEM

The Urdhva-Tiryak Sutra (the Criss-Cross system) is by far the most popular system of squaring numbers amongst practitioners of Vedic Mathematics. The reason for its popularity is that it can be used for any type of numbers.

Ex: Find the square of 23.

Ans: (a) * *

(b) * *
     X
     * *

(c) * *

    * *

      2   3
   × 2   3

(a) First, we multiply 3 by 3 and get the answer as 9.
    (Answer at this stage is _____9).

(b) Next, we cross multiply (2 × 3) and add it with (2 × 3).
    The final answer is 12. We write down 2 and carry over
    1.
    (Answer at this stage is _____29).

(c) Thirdly, we multiply (2 × 2) and add the 1 to it. The
    answer is 5.
    The final answer is 529.

Similarly, numbers of higher orders can be squared. Refer to
the chapter on Criss-Cross system for further reference.

(B) FORMULA METHOD

There are various formulae used in general mathematics to
square numbers instantly. Let us discuss them one by one.

$$(i) \quad (a + b)^2 = a^2 + 2ab + b^2$$

This method is generally to square numbers which are near
multiples of 10. In this method, a given number is expanded in
such a manner that the value of 'a' is a number which can be
easily squared and the value of 'b' is a small number which too
can be easily squared.

(Q) Find the square of 1009.

We represent the number 1009 as 1000 + 9. Thus, we have converted it into a form of (a + b) where the value of a is 1000 and the value of b is 9.

$$(1000 + 9)^2 = (1000)^2 + 2 (1000) (9) + (9)^2$$
$$= 1000000 + 18000 + 81$$
$$= 1018081$$

(Q) Find the square of 511.

The number 511 will be written as 500 + 11

$$(500 + 11)^2 = (500)^2 + 2 (500) (11) + (11)^2$$
$$= 250000 + 11000 + 121$$
$$= 261121$$

The second formula that we will discuss is also very well known to the students. It is a part of the regular school curriculum. This formula is used to square numbers which can be easily expressed as a difference of two numbers 'a' and 'b' in such a way that the number 'a' is one which can be easily squared and the number 'b' is a small number which too can be easily squared. The formula is,

$$(ii) \ (a - b)^2 = a^2 - 2ab + b^2$$

This formula is very much like the first one. The only difference is that the middle term carries a negative sign in this formula.

(Q) Find the square of 995.

We will express the number 995 as (1000 – 5)

$$(1000 - 5)^2 = (1000)^2 - 2 (1000) (5) + (5)^2$$
$$= 1000000 - 10000 + 25$$
$$= 990025$$

(Q) Find the square of 698.

We will express the number 698 as (700 - 2)

$$(700 - 2)^2 \quad = (700)^2 - 2\,(700)\,(2) + (2)^2$$
$$= 490000 - 2800 + 4$$
$$= 487204$$

           *   *   *   *   *   *

Thus, we see that the two formulae can help us find the squares of any number above and below a round figure respectively. There is another formulae which is used to find the square of numbers, but it is not so popular. I discuss it below.

We know that:

$$a^2 - b^2 = (a + b)\,(a - b)$$
$$\text{(Therefore) } a^2 = (a + b)\,(a - b) + b^2$$

This is the formula that we will be using: $\mathbf{a^2 = (a + b)\,(a - b) + b^2}$

METHOD

Suppose we are asked to find the square of a number. Let us call this number 'a'. Now in this case we will use another number 'b' in such a way that either $(a + b)$ or $(a - b)$ can be easily squared.

(Q) Find the square of 72.

Ans: In this case, the value of 'a' is 72. Now, we know that

$$a^2 = (a + b)\,(a - b) + b^2$$

Substituting the value of 'a' as 72, we can write the above formula as:

$$(72)^2 = (72 + b)\,(72 - b) + b^2$$

We have substituted the value of 'a' as 72. However, we cannot solve this equation because a variable 'b' is still present. Now, we have to substitute the value of 'b' with such a number that the whole equation becomes easy to solve.

Let us suppose I take the value of 'b' as 2.

Then the equation becomes,

$(72)^2 = (72 + 2) (72 - 2) + (2)^2$
$\quad\quad = (74) (70) + 4$

In this case we can find the answer by multiplying 74 by 70 and adding 4 to it. However, if one finds multiplying 74 by 70 difficult, we can simplify it still further. First, multiply 70 by 70 and then multiply 4 by 70 and add both for the answer.

Let us continue the example given above:

$= (70 \times 70) + (4 \times 70) + 4$
$= 4900 + 280 + 4$
$= \mathbf{5184}$

Thus, the square of 72 is 5184.

In this example we have taken the value of 'b' as 2. Because of this, the value of (a - b) became 70 and the multiplication procedure became easy (as the number 70 ends with a zero).

(Q) Find the square of 53.

Ans: Uing the formula $a^2 = (a + b) (a - b) + b^2$ and taking the value of 'a' as 53, we have:

$(53)^2 = (53 + b) (53 - b) + (b)^2$

Now we have to find a suitable value for 'b'. If we take the value of 'b' as 3, the expression (53 - 3) will be 50 and hence it will simplify the multiplication procedure. So we will take the value of 'b' as 3 and the equation will become:

$(53)^2 = (53 + 3) (53 - 3) + (3)^2$
$\quad\quad = (56) (50) + 9$
$\quad\quad = (50 \times 50) + (6 \times 50) + 9$
$\quad\quad = 2500 + 300 + 9$
$\quad\quad = \mathbf{2809}$

(Q) Find the square of 67.

Ans: In this case the value of 'a' is 67. Next, we will substitute 'b' with a suitable value. In this case, let us take the value of 'b' as 3 so that the value of (a + b) will become (67 + 3) which equals 70.

Thus: $(67)^2$ = (67 + 3) (67 - 3) + $(3)^2$
       = (70) (64) + 9
       = (70 × 60) + (70 × 4) + 9
       = 4200 + 280 + 9
       = **4489**

(Q) Find the square of 107.

Ans: In this case, we will take the value of 'a' as 107 and take the value of 'b' as 7. The equation becomes:

$(107)^2$ = (107 + 7) (107 - 7) + $(7)^2$
         = (114) (100) + 49
         = 11400 + 49
         = 11449

(Q) Find the square of 94.

Ans: In this example we will take the value of 'a' as 94. Next, we will take the value of 'b' as 6 so that the value of (a + b) becomes 100.

$(94)^2$ = (94 + 6) (94 - 6) + $(6)^2$
        = (100) (88) + 36
        = 8836

EXERCISE

PART A

Q. (1)   Find the squares of the following numbers using the Criss-Cross System.

    (1) 42
    (2) 33
    (3) 115

PART B

Q. (2)   Find the squares of the numbers using the formula for $(a + b)^2$.

    (1) 205
    (2) 2005
    (3) 4050

Q. (3)   Find the squares of the numbers using the formula for $(a - b)^2$.

    (4) 9991
    (5) 9800
    (6) 1090

PART C

Q. (4)   Find the squares of the following numbers using the formula: $a^2 = (a + b)(a - b) + b^2$.

    (1) 82
    (2) 49
    (3) 109
    (4) 97

Cube Roots of Perfect Cubes

In problems of algebra dealing with factorization, equations of third power and in many problems of geometry related to three dimensional figures, one will often find the need to calculate the cube root of numbers. Calculating the cube root of a number by the traditional method is a slightly cumbersome procedure, but the technique used by Vedic Mathematicians is so fast that one can get the answer in two to three seconds!

The technique for solving cube roots is simply so amazing that the student will be able to correctly predict the cube root of a number just by looking at it and without the need for any intermediate steps.

You might find it difficult to believe, but at the end of this study, you will be calculating cube roots of complicated numbers like 262144, 12167 and 117649 in 2-3 seconds. Even primary school students who have learnt these techniques from me are able to calculate cube-roots in a matter of seconds.

Before we delve deeper in this study, let us clear our concepts relating to cube roots.

WHAT IS CUBE ROOTING?

Let us take the number 3. When we multiply 3 by itself we are said to have squared the number 3. Thus 3×3 is 9. When we multiply 4 by itself we are said to have squared 4 and thus 16 is the square of 4.

Similarly the square of 5 is 25 (represented as 5^2)
The square of 6 is 36 (represented as 6^2)

In squaring, we multiply a number by itself, but in cubing we multiply a number by itself and then multiply the answer by the original number once again.

Thus, the cube of 2 is $2 \times 2 \times 2$ and the answer is 8. (represented as 2^3)

The cube of 3 is $3 \times 3 \times 3$ and the answer is 27. (represented as 3^3)

Basically, in squaring we multiply a number by itself and in cubing we multiply a number twice by itself.

Now, since you have understood what cubing is it will be easy to understand what cube rooting is. Cube-rooting is the procedure of determining the number which has been twice multiplied by itself to obtain the cube. Calculating the cube-root is the reciprocal procedure of calculating a cube.

Thus, if 8 is the cube of 2, then 2 is the cube-root of 8.

If 27 is the cube of 3 then 3 is the cube root of 27 and so on.

In this chapter, we will learn how to calculate cube-roots. Thus, if you are given the number 8 you will have to arrive at the number 2. If you are given the number 27 you will have to arrive at 3. However, these are very basic examples. We shall be cracking higher order numbers like 704969, 175616, etc.

At this point, I would like to make a note that the technique provided in this chapter can be used to find the cube-roots of perfect cubes only. It cannot be used to find the cube root of imperfect cubes.

METHOD

I have given below a list containing the numbers from 1 to 10 and their cubes. This list will be used for calculating the cube roots of higher order numbers. With the knowledge of these numbers, we shall be able to solve the cube-roots instantly. Hence, I urge the reader to memorize the list given below before proceeding ahead with the chapter.

NUMBER	CUBE
<u>1</u>	<u>1</u>
<u>2</u>	<u>8</u>
<u>3</u>	2<u>7</u>
<u>4</u>	6<u>4</u>
<u>5</u>	12<u>5</u>
<u>6</u>	21<u>6</u>
<u>7</u>	34<u>3</u>
<u>8</u>	51<u>2</u>
<u>9</u>	72<u>9</u>
1<u>0</u>	100<u>0</u>

The cube of 1 is 1, the cube of 2 is 8, the cube of 3 is 27 and so on....

Once you have memorized the list I would like to draw your attention to the underlined numbers in the key. You will notice that I have underlined certain numbers in the key. These underlined numbers have a unique relationship amongst themselves.

In the first row, the underlined numbers are 1 and 1. It establishes a certain relationship that if the last digit of the cube is 1 then the last digit of the cube root is also 1.

In the second row, the underlined numbers are 2 and 8. It establishes a relationship that if the last digit of the cube is 8 then the last digit of the cube root is 2. Thus, in any given cube if the last digit of the number is 8 the last digit of its cube-root will always be 2.

In the third row, the underlined numbers are 3 and 7 (out of 27 we are interested in the last digit only and hence we have underlined only 7). We can thus conclude that if the last digit of a cube is 7 the last digit of the cube root is 3.

And like this if we observe the last row where the last digit of 10 is 0 and the last digit of 1000 is also 0. Thus, when a cube ends in 0 the cube-root also ends in 0.

On the basis of the above observations, we can form a table as given below:

The Last Digit Of The Cube	The Last Digit Of The Cube Root
1	1
2	8
3	7
4	4
5	5
6	6
7	3
8	2
9	9
0	0

From the above table, we can conclude that all cube-roots end with the same number as their corresponding cubes except for pairs of 3 & 7 and 8 & 2 which end with each other.

There is one more thing to be kept in mind before solving cube-roots:

Whenever a cube is given to you to calculate its cube-root, you must put a slash before the last three digits.

If the cube given to you is 103823 you will represent it as 103|823

If the cube given to you is 39304, you will represent it as 39|304

Immaterial of the number of digits in the cube, you will always put a slash before the last three digits.

SOLVING CUBE ROOTS

No.	1	2	3	4	5	6	7	8	9	10
Cube	1	8	27	64	125	216	343	512	729	1000

We will be solving the cube root in 2 parts. First, we shall solve the right hand part of the answer and then we shall solve the left hand part of the answer. If you wish you can solve the left hand part before the right hand part. There is no restriction on either method but generally people prefer to solve the right hand part first.

As illustrative examples, we shall take four different cubes.

(Q) Find the cube root of 287496.

- We shall represent the number 287496 as 287|496

- Next, we observe that the cube 287496 ends with a 6 and we know that when the cube ends with a 6, the cube root also ends with a 6. Thus our answer at this stage is ___6. We have thus got the right hand part of our answer.

- To find the left hand part of the answer we take the number which lies to the left of the slash. In this case, the number lying to the left of the slash is 287. Now, we need to find two perfect cubes between which the number 287 lies in the number line. From the key, we find that 287 lies between the perfect cubes 216 (**the cube of 6**) and 343 (**the cube of 7**).

- Now, out of the numbers obtained above, we take the smaller number and put it on the left hand part of the answer. Thus, out of 6 and 7, we take the smaller number 6 and put it beside the answer of ___6 already obtained. Our final answer is 66. Thus, 66 is the cube root of 287496.

(Q) Find the cube root of 205379.

- We represent 205379 as 205 379
- The cube ends with a 9, so the cube root also ends with a 9. (The answer at this stage is _____9.
- The part to the left of the slash is 205. It lies between the perfect cubes 125 (the cube of 5) and 216 (the cube of 6)
- Out of 5 and 6, the smaller number is 5 and so we take it as the left part of the answer. The final answer is 59.

(Q) Find the cube root of 681472.

- We represent 681472 as 681 472
- The cube ends with a 2, so the root ends with an 8. The answer at this stage is _____8.
- 681 lies between 512 (the cube of 8) and 729 (the cube of 9).
- The smaller number is 8 and hence our final answer is 88.

(Q) Find the cube root of 830584.

- The cube ends with a 4 and the root will also end with a 4.
- 830 lies between 729 (the cube of 9) and 1000 (the cube of 10).
- Since, the smaller number is 9, the final answer is 94.

You will observe that as we proceeded with the examples, we took much less time to solve the cube-roots. After some practice you will be able to solve the cube-roots by a mere observation of the cube and without the necessity of doing any intermediary steps.

It must be noted that immaterial of the number of digits in the cube, the procedure for solving them is the same.

(Q) Find the cube root of 2197.

- The number 2197 will be represented as 2 197
- The cube ends in 7 and so the cube root will end with a 3.

We will put 3 as the right hand part of the answer.

- The number 2 lies between 1 (the cube of 1) and 8 (the cube of 2).

- The smaller number is 1 which we will put as the left hand part of the answer. The final answer is 13.

We may thus conclude that there exists only one common procedure for solving all types of perfect cube-roots.

In my seminars, the participants often ask what is the procedure of solving cube roots of numbers having more than 6 digits. (All the examples that we have solved before had 6 or fewer digits.)

Well the answer to this question is that the procedure for solving the problem is the same. The only difference in this case is that you will be expanding the number line.

Let us take an example.

We know that the cube of 9 is 729 and the cube of 10 is 1000. Now let us go a step ahead and include the higher numbers. We know that the cube of 11 is 1331 and the cube of 12 is 1728.

Number	9	10	11	12
Cube	729	1000	1331	1728

(Q) Find the cube root of 1157625.

- We put a slash before the last three digits and represent the number as 1157| 625.

- The number 1157625 ends with a 5 and so the root also ends with a 5. The answer at this stage is _____5.

- We take the number to the left of the slash, which is 1157. In the number line it lies between 1000 (the cube of 10) and 1331 (the cube of 11).

- Out of 10 and 11, we take the smaller number 10 and put it beside the 5 already obtained. Our final answer is 105.

(Q) Find the cube root of 1404928.

- The number will be represented as 1404 928
- The number ends with a 8 and so the cube-root will end with a 2.
- 1404 lies between 1331 (the cube of 11) and 1728 (the cube of 12). Out of 11 and 12 the smaller number is 11 which we will put beside the 2 already obtained. Hence, the final answer is 112.

The two examples mentioned above were just for explanation purposes. Under normal circumstances, you will be asked to deal with cubes of 6 or less than 6 digits in your exams. Hence, knowledge of the key which contains cubes of numbers from 1 to 10 is more than sufficient. However, since we have dealt with advanced level problems also, you are well equipped to deal with any kind of situation.

COMPARISON

As usual, we will be comparing the normal technique of calculation with the Vedic technique. In the traditional method of calculating cube-roots we use prime numbers as divisors.

Prime numbers include numbers like 2, 3, 5, 7, 11, 13 and so on.

Let us say you want to find the cube root of 64. Then, the process of calculating the cube root of 64 is as explained below.

2	64
2	32
2	16
2	8
2	4
2	2
	1

First, we divide the number 64 by 2 and get the answer 32.

- 32 divided by 2 gives 16
- 16 divided by 2 gives 8
- 8 divided by 2 gives 4
- 4 divided by 2 gives 2
- 2 divided by 2 gives 1

(We terminate the division when we obtain 1)

Thus, $64 = 2 \times 2 \times 2 \times 2 \times 2 \times 2$

To obtain the cube root, for every three similar numbers we take one number. So, for the first three 2's, we take one 2 and for the next three 2's we take one more 2.

When these two 2's are multiplied with each other we get the answer 4 which is the cube root of 64.

It can be represented as:

$$64 = \underline{2 \times 2 \times 2} \times \underline{2 \times 2 \times 2}$$

$$\downarrow \downarrow$$

$$ 2 \quad \times \quad 2 \qquad \text{equals} \quad 4.$$

Hence, 4 is the cube root of 64.

Similarly, to find the cube root of 3375 by the traditional method, we can use the following procedure.

5	3375
5	675
5	135
3	27
3	9
3	3
	1

$$3375 = \underline{5 \times 5 \times 5} \times \underline{3 \times 3 \times 3}$$

$$\downarrow \downarrow$$

$$ 5 \quad \times \quad 3 \qquad = 15$$

Hence, 15 is the cube-root of 3375.

After studying the above two examples, the reader will agree with me that the traditional method is cumbersome and time-consuming compared to the method used by Vedic mathematicians. However, you will be shocked to see the difference between the two methods when we try to calculate the cube root of some complicated number.

Example: Find the cube root of 262144.

The Traditional Method   The Vedic Mathematics method

2	262144
2	131072
2	65536
2	32768
2	16384
2	8192
2	4096
2	2048
2	1024
2	512
2	256
2	128
2	64
2	32
2	16
2	8
2	4
2	2
	1

$262 \mid 144 = \underline{64}$

$$= 2 \times 2 \times 2 \times 2 \times 2 \times 2$$
$$= 64$$

It can be observed from the comparison that not only does the Vedic method helps us to find the answer in one line but also helps us to find the answer directly without the need for any intermediate steps. This characteristic of this system helps students in instantly cracking such problems in competitive exams.

With this comparison we terminate this study. Students are urged to solve the practice exercise before proceeding to the next chapter.

EXERCISE

PART A

Q. (1) Find the cube roots of the following numbers **with** the aid of writing material.

(1) 970299

(2) 658503

(3) 314432

(4) 110592

(5) 46656

(6) 5832

(7) 421875

(8) 1030301

PART B

Q. (2) Find the cube roots of the following numbers **without** the aid of writing material.

(1) 132651	(5) 474552
(2) 238328	(6) 24389
(3) 250047	(7) 32768
(4) 941192	(8) 9261

Square Roots of Perfect Squares

We have divided the study of square roots in two parts. In this chapter we will study how to find the square roots of perfect squares; in the 'Advance Level' we will study the square roots of general numbers. Most school and college exams ask the square roots of perfect squares. Therefore, this chapter is very useful to students giving such exams. Students of higher classes and other researchers will find the thirteenth chapter useful as they will be able to study the Vedic Mathematics approach to calculate square roots of any given number — perfect as well as imperfect.

The need to find perfect square roots arises in solving linear equations, quadratic equations and factorizing equations. Solving square roots is also useful in geometry while dealing with the area, perimeter, etc. of geometric figures. The concepts of this chapter will also be useful in dealing with the applications of the Theorem of Pythagoras.

The technique of finding square roots of perfect squares is similar to the technique of finding the cube root of perfect cubes. However, the former has an additional step and hence it is discussed after having dealt with cube roots.

WHAT IS SQUARE ROOT

To understand square roots it will be important to understand what are squares. Squaring of a number can be defined as multiplying a number by itself. Thus, when we multiply 4 by 4 we are said to have 'squared' the number four.

The symbol of square is represented by putting a small 2 above the number.

E.g. (a) $4^2 = 4 \times 4 = 16$
    (b) $5^2 = 5 \times 5 = 25$

From the above example we can say that 16 is the square of 4, and 4 is the **'square root'** of 16. Similarly, 25 is the square of 5, and 5 is the **square-root** of 25.

METHOD

To find the square roots it is necessary to be well versed with the squares of the numbers from 1 to 10. The squares are given below. Memorize them before proceeding ahead.

NUMBER	SQUARE
1	1
2	4
3	9
4	16
5	25
6	36
7	49
8	64
9	81
10	100

In the chapter dealing with perfect cube roots we observed that if the last digit of the cube is 1 the last digit of the cube root is also 1. If the last digit of the cube is 2 then the last digit of the cube-root is 8 and so on. Thus, for every number there was a unique corresponding number.

However in square roots we have more than one possibility for every number. Look at the first row. Here, we have 1 in the number column and 1 in the square column. Similarly, in the ninth row we have 1 in the number column and 1 (of 81) in the square column. Thus, if the number ends in 1, the square root ends in 1 or 9 (because 1 × 1 is **one** and 9 × 9 is eighty-**one**). Do not worry if you do not follow this immediately. You may glance at the table below as you read these explanations, then all will be clear.

- Similar to the 1 and 9 relationship, if a number ends in 4 the square root ends in 2 or 8. (because 2 × 2 is **four** and 8 × 8 is sixty-**four**)
- If a number ends in 9, the square root ends in 3 or 7. (because 3 × 3 is **nine** and 7 × 7 is forty-**nine**)
- If a number ends in 6, the square root ends in 4 or 6. (because 4 × 4 is 1**6** and 6 × 6 is 3**6**)
- If the number ends in 5, the square root ends in 5 (because 5 × 5 is twenty-**five**)
- If the number ends in 0, the square root also ends in 0 (because 10 × 10 is 1**00**)

On the basis of such observations, we can form a table as given below:

The Last Digit of the Square	The Last Digit of the Square Root
1	1 or 9
4	2 or 8
9	3 or 7
6	4 or 6
5	5
0	0

Whenever we come across a square whose last digit is 9, we can conclude that the last digit of the square root will be 3 or 7. Similarly, whenever we come across a square whose last digit is 6, we can conclude that the last digit of the square root will be 4 or 6 and so on....

Now, I want you to look at the column in the left. It reads 'Last digit of the square' and the numbers contained in the column are 1,4, 9, 6, 5, and 0. Note that the numbers 2, 3, 7 and 8 are absent in the column. That means there is no perfect square which ends with the numbers 2, 3, 7 or 8. Thus we can deduct a rule:

'A perfect square will never end with the digits 2, 3, 7 or 8'

At this point we have well understood how to find the last digit of a square root. However, in many cases we will have two possibilities out of which one is correct. Further, we do not know how to find the remaining digits of the square root. So we will solve a few examples and observe the technique used to find the complete square root.

Before proceeding ahead with the examples, I have given below a list of the squares of numbers which are multiples of 10 up to 100. This table will help us to easily determine the square roots.

NUMBER	SQUARE
10	100
20	400
30	900
40	1600
50	2500
60	3600
70	4900
80	6400
90	8100
100	10000

(Q) Find the square root of 7744.

- The number 7744 ends with 4. Therefore the square root ends with 2 or 8. The answer at this stage is __2 or __8.

- Next, we take the complete number 7744. We find that the number 7744 lies between 6400 (which is the square of 80) and 8100 (which is the square of 90).

$$70 - 4900$$
$$80 - 6400$$
$$\qquad\qquad\diagdown$$
$$\qquad\qquad\qquad > 7744$$
$$90 - 8100 \diagup$$
$$100 - 10000$$

The number 7744 lies between 6400 and 8100. Therefore, the square root of 7744 lies between the numbers 80 and 90.

- From the first step we know that the square root ends with 2 or 8. From the second step we know that the square root lies between 80 and 90. Of all the numbers between 80 and 90 (81, 82, 83, 84, 85, 86, 87, 88, 89) the only numbers ending with 2 or 8 are 82 or 88. Thus, out of 82 or 88, one is the correct answer.

  (Answer at this stage is 82 or 88).

- Observe the number 7744 as given below:

$$80 - 6400$$
$$\qquad\qquad\diagdown$$
$$\qquad\qquad\qquad > 7744$$
$$90 - 8100 \diagup$$

Is it closer to the smaller number 6400 or closer to the bigger number 8100 ?

If the number 7744 is closer to the smaller number 6400 then take the smaller number 82 as the answer. However, if it is closer to the bigger number 8100, then take 88 as the answer.

In this case, we observe that 7744 is closer to the bigger number 8100 and hence we take 88 as the answer.

The square root of 7744 is 88.

(Q) Find the square root of 9801.

- The last digit of the number 9801 is 1 and therefore the last digit of the square root will be either 1 or 9. The answer at this stage is _____1 or _____9.

- Next, we observe that the number 9801 lies between 8100 (which is the square of 90) and 10000 (which is the square of 100). Thus, our answer lies between 90 and 100. Our possibilities at this stage are:

  91, 92, 93, 94, 95, 96, 97, 98, 99

- However, from the first step we know that the number ends with a 1 or 9. So, we can eliminate the numbers that do not end with a 1 or 9.

  91, 92, 93, 94, 95, 96, 97, 98, **99**

- The two possibilities at this stage are 91 or 99. Lastly, we know that the number 9801 is closer to the bigger number 10,000 and so we take the bigger number 99 as the answer.

$$90 - 8100$$
$$\searrow$$
$$\qquad\qquad 9801$$
$$\nearrow$$
$$100 - 10000$$

(Q) Find the square root of 5184.

- 5184 ends in 4. So the square root ends in either 2 or 8 (Answer = _____2 or _____8)

- 5184 is between 4900 and 6400. So the square root is between 70 and 80. Combining the first two steps, the only two possibilities are 72 and 78

- Out of 4900 and 6400, our number 5184 is closer to the smaller number 4900 (70×70). Thus, we take the smaller number 72 as the correct answer

(Q) Find the square root of 2304.

- 2304 ends in 4 and so the root either ends in a 2 or in a 8

- 2304 lies between 1600 and 2500. So, the root lies between 40 and 50.

- Thus, the two possibilities are 42 and 48.
- Lastly, the number 2304 is closer to the bigger number 2500. Hence, out of 42 and 48 we take the bigger number 48 as the correct answer.

(Q) Find the square root of 529.

- 529 ends with a 9. The answer is ____3 or _____7.
- It lies between 20 and 30. The possibilities are 23 or 27.
- 529 is closer to the smaller number 400 and hence 23 is the answer.

We have seen five different examples and calculated their square roots. However, the final answer in each case is always a two-digit number. In most exams and even in general life, one will come across squares whose roots are a two-digit answer. Thus, the above examples are sufficient and there is no necessity to stretch the concept further. However, we will study a couple of examples involving big numbers so that you will understand the fundamentals thoroughly.

(Q) Find the square root of 12544.

The number 12544 ends with a 4. So, the square root ends with 2 or 8. The answer at this stage is __2 or __8.

- Further, we know that the square of 11 is 121 and so the square of 110 is 12100. Similarly, the square of 12 is 144 and so the square of 120 is 14400.

$$
\begin{aligned}
90 &- 8100 \\
100 &- 10000 \\
110 &- 12100 \\
& \rangle 12544 \\
120 &- 14400
\end{aligned}
$$

- The number 12544 lies between 12100 (which is the square of 110) and 14400 (which is the square of 120). Thus, the square root of 12544 lies between 110 and 120.
- But we know that the square root ends with 2 or 8. Hence, our only possibilities are 112 or 118.

- Lastly, 12544 is closer to the smaller number 110 and hence we take the smaller possibility 112 as the answer.

The square root of 12544 is 112.

(Q)   Find the square root of 25281.

- The number 25281 ends with a 1. Therefore the square root ends with a 1 or a 9. The answer at this stage is ____1 or ____9.

- We know that the square of 15 is 225 and therefore the square of 150 is 22500. Similarly, the square of 16 is 256 and therefore the square of 160 is 25600.

$$150 - 22500$$
$$\searrow$$
$$25281$$
$$160 - 25600\nearrow$$

- We know that the root lies between 150 and 160 and hence the only possibilities are 151 and 159.

- Lastly, the number 25281 is closer to the bigger number 25600 and hence we take the bigger number 159 as the correct answer.

We have thus seen that the concept can be expanded to numbers of any length.

COMPARISON

As usual, we will be comparing the normal technique of calculation with our approach. In the traditional method of calculating square-roots we use prime numbers as divisors.

Prime numbers are numbers which can be divided by themselves and by 1 only. They will not come in the multiplication table of any other number. They include numbers like 2, 3, 5, 7, 11, 13 and so on.

Let us say you want to find the square root of 256. Then, the process of calculating the square root of 256 is as explained below.

2	256
2	128
2	64
2	32
2	16
2	8
2	4
2	2
	1

First we divide the given number 256 by the prime number 2 and get the answer as 128.

- Next, we divide 128 by 2 and get the answer 64
- 64 divided by 2 gives 32
- 32 divided by 2 gives 16
- 16 divided by 2 gives 8
- 8 divided by 2 gives 4
- 4 divided by 2 gives 2
- 2 divided by 2 gives 1

  (We terminate the division when we obtain 1).

Thus, $256 = 2 \times 2 \times 2 \times 2 \times 2 \times 2 \times 2 \times 2$.

To obtain the square root, for every two similar numbers, we take one number. So, we form four pairs containing two 2's each. From every pair we take one 2.

It can be represented as:

$$256 = \underline{2 \times 2} \times \underline{2 \times 2} \times \underline{2 \times 2} \times \underline{2 \times 2}$$
$$\quad\quad\quad \downarrow \quad\quad\quad \downarrow \quad\quad\quad \downarrow \quad\quad\quad \downarrow$$
$$\quad\quad\quad 2 \quad \times \quad 2 \quad \times \quad 2 \quad \times \quad 2 \quad = 16$$

Thus, the square root of 256 is 16.

(Q) Find the square root of 196 using prime factors.

```
2 | 196
2 | 98
7 | 49
7 | 7
  | 1
```

$196 = 2 \times 2 \times 7 \times 7$

$\qquad\qquad\quad 2 \quad \times \quad 7 \quad = 14$

Therefore, the square root of 196 is 14.

From the above two examples it is clear that the prime factor method of calculating square roots is time consuming and tedious. Further, if small numbers like 256 and 196 take such a lot of time, one can imagine how difficult it will be to calculate the square roots of numbers like 8281, 7744, etc. Some people find it simply impossible to calculate the square roots of such numbers using the prime factor technique. Hence, the alternate approach as mentioned in this chapter will be of immense utility to the student.

(Q) Calculate the square root of 576.

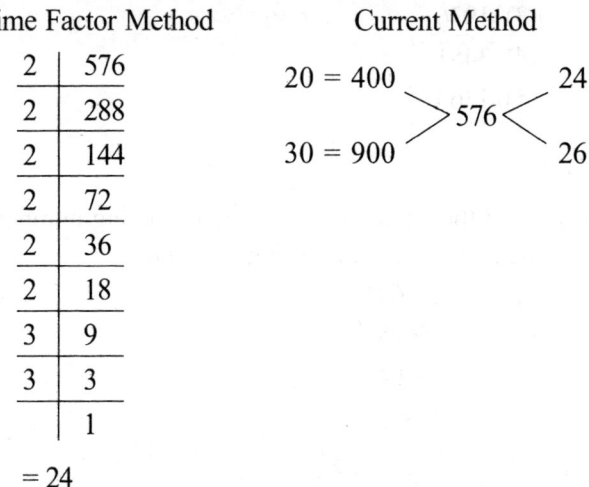

Prime Factor Method   Current Method

```
2 | 576
2 | 288
2 | 144
2 | 72
2 | 36
2 | 18
3 | 9
3 | 3
  | 1
```

= 24

$20 = 400$

$30 = 900$

$> 576 <$

24

26

From the comparison we can see that the method described
in this chapter is much faster and the chances of making a
mistake are greatly reduced. Further, while the prime factor
method will prove extremely tedious to calculate the square
roots of numbers like 4356 and 6561, the current method will
help us calculate them instantly!

EXERCISE

PART A

Q. (1) Find the square roots of the following numbers with the
aid of writing material.
(1) 9216
(2) 7569
(3) 5329
(4) 3364
(5) 1681

PART B

Q. (1) Find the square roots of the following numbers without
the aid of writing material.
(1) 9801
(2) 5625
(3) 1936
(4) 3481
(5) 1369

PART C

Q. (1) Find the square roots of the following numbers with or
without the aid of writing material.
(1)   12769
(2)   15625
(3)   23104
(4)   11881

INTERMEDIATE LEVEL

Base Method of Multiplication

The Base Method of multiplication is a wonderful contribution of Vedic Mathematics. The name Base Method is given by practitioners of Vedic Mathematics in western countries. However, the actual Sanskrit sutra as given by Swamiji to define this system is:

'Nikhilam Navatascaramam Dasatah.' It means 'all from 9 and the last from 10.'

For all practical purposes, we shall be calling the system elaborated in this chapter as the Nikhilam method or simply the 'Base Method.'

This method is used to multiply numbers. It is of immense help in certain cases where traditional multiplication takes a long time to calculate the answer. Let us take the case of multiplying the number 9999999 by 9999998. If you go by the traditional method it will take a long time to multiply the numbers and calculate the product. However with the technique described in the Base Method one can find the answer in less than 5 seconds.

The study of the Base Method is crucial to understand the other formulae of Vedic Mathematics. There is a corollary of the Base Method which is called the Yavadunam Rule. This sutra is used in squaring numbers and is discussed in the next chapter.

OVERVIEW

This system is called the Base Method because in this system we use a certain number as a base. This base can be any number, but generally we use powers of 10. The powers of 10 include numbers like 10, 100, 1000, 10000, etc. We select a particular base depending upon the numbers given in the question. Suppose we are asked to multiply 97 by 99; in this case the appropriate base would be 100 as both these numbers are closer to 100. If we are asked to multiply 1005 by 1020 then the appropriate base would be 1000 as both the numbers are closer to 1000.

Secondly, we will find the answer in two parts — the left hand side and the right hand side. The left hand side will be denoted by the acronym LHS and the right hand side will be denoted by the acronym RHS.

Let us have a look at the procedure involved in this technique of multiplication. I have outlined below the four steps required in this technique.

STEPS

  (a) Find the Base and the Difference
  (b) Number of digits on the RHS = Number of zeros in the base
  (c) Multiply the differences on the RHS
  (d) Put the Cross Answer on the LHS

These are the four primary steps that we will use in any given problem. I urge the reader to read the steps 2-3 times before proceeding ahead. To understand how the system works we will solve three different questions simultaneously.

(Q) Find the answer to the following questions:

(A)	(B)	(C)
9 7	9 9 8 9	9 9 9 9 9 9 9
× 9 9	× 9 9 9 5	× 9 9 9 9 9 9 9

STEP A: Find the Base and the difference.

The first part of the step is to find the base. Have a look at example A. In this example the numbers are 97 and 99. We know that we can take only powers of 10 as bases. The powers of 10 are numbers like 10, 100, 1000, 10000, etc. In this case since both the numbers are closer to 100 we will take 100 as the base.

Similarly, in example B both the numbers are closer to 10,000 and so we will take 10,000 as the base.

In example C, both the numbers are closer to 1,00,00,000 (one crore) and hence we will take one crore as the base.

So, we have found the bases.

(A)	(B)	(C)
100	**10,000**	**1,00,00,000**
9 7 – 3	9 9 8 9 – 1 1	9 9 9 9 9 9 9 – 1
9 9 – 1	9 9 9 5 – 5	9 9 9 9 9 9 9 – 1

We are still on step A. Next, we have to find the differences.

In Example A the difference between 100 and 97 is 3 and the difference between 100 and 99 is 1.

In Example B the difference between 10,000 and 9989 is 11 and the difference between 10,000 and 9995 is 5.

In Example C, the difference of both the numbers from the base is 1.

Step A is complete. The question looks as given below:

(A)	(B)	(C)
100	**10,000**	**1,00,00,000**
9 7 – 3	9 9 8 9 – 11	9 9 9 9 9 9 9 – 1
9 9 – 1	9 9 9 5 – 5	9 9 9 9 9 9 9 – 1

Step B: Number of digits in RHS = No. of zeros in the base.

Now, we come to step B. In this step we are going to find the answer of the multiplication question in two parts, viz. the left hand side or LHS and the right hand side or the RHS. Step B says that the number of digits to be filled in the Right Hand Side of the answer should be equal to the number of zeros in the base.

In example A the base 100 has two zeros. Hence, the RHS will be filled in by a two-digit number.

In example B, the base 10,000 has four zeros and hence the RHS will be filled by a four-digit number.

In example C the base one crore has seven digits and hence the RHS will be filled in by a seven-digit number.

Let us make provisions for the same:

(A)	(B)	(C)
100	**10,000**	**1,00,00,000**
9 7 – 3	9 9 8 9 – 11	9 9 9 9 9 9 9 – 1
9 9 – 1	9 9 9 5 – 5	9 9 9 9 9 9 9 – 1
├ –	├ – – –	├ – – – – – –

We have separated the LHS and the RHS with a straight line. The RHS will have as many digits as the number of zeros in the base and so we have put empty blanks in the RHS of equal number.

Step C: Multiply the differences in RHS.

The third step (step C) says to multiply the differences and write the answer in the right-hand side.

In example A we multiply the differences, viz. -3 by -1 and get the answer as 3. However, the RHS has to be filled by a two-digit number. Hence, we convert the answer 3 into 03 and put it on the RHS.

In example B we multiply -11 by -5 and get the answer as 55. Next, we convert 55 into a four-digit number 0055 and put it on the RHS.

In example C, we multiply the differences -1 by -1 and get the answer as 1. We convert it into a seven-digit number and put it on the RHS as 0000001.

(A)	(B)	(C)
100	**10,000**	**1,00,00,000**
$9\ 7 - 3$	$9\ 9\ 8\ 9 - 11$	$9\ 9\ 9\ 9\ 9\ 9\ 9 - 1$
$9\ 9 - 1$	$9\ 9\ 9\ 5 - 5$	$9\ 9\ 9\ 9\ 9\ 9\ 9 - 1$
|03	|0055	|0000001

STEP D: Put the cross answer in the LHS.

Now we come to the last step. At this stage we already have the right-hand part of the answer. If you are giving any competitive exam and the right-hand part of the answer uniquely matches with one of the given alternatives, you can straight-away tick that alternative as the correct answer. However, the multiplication in our case is still not complete. We still have to get the left hand side of the answer.

Step D says to put the cross answer in the left hand side. Let us observe how Step D will be applied in each of the alternatives.

In example A, the cross answer can be obtained by doing $(97 - 1)$ or $(99 - 3)$. In either case the answer will be 96. This 96 we will put on the LHS. But we already had 03 on the

RHS. Hence, the complete answer is 9603.

$$9\,7 - 3$$
$$\times$$
$$9\,9 - 1$$

In example B, we subtract (9989 – 5) and get the answer as 9984. We can even subtract (9995 – 11). In either case, the answer is the same. We put 9984 on the LHS. We already had 0055 in the RHS. The complete answer is 99840055.

In example C, we cross subtract 1 from 9999999. The answer we get is 999998 which we put on the LHS. The final answer is 99999980000001.

Thus, our multiplication procedure is complete.

(A)	(B)	(C)
100	**10,000**	**1,00,00,000**
9 7 – 3	9 9 8 9 – 1 1	9 9 9 9 9 9 9 – 1
9 9 – 1	9 9 9 5 – 5	9 9 9 9 9 9 9 – 1
96│03	9984│0055	9999998│0000001

At first glance, this system might appear too cumbersome and lengthy. In fact, we have taken more than three pages to solve three simple examples. However, because I was explaining the technique for the first time, I elaborated every single step and provided explanation for it. Thus, it appears lengthy. In reality it is not so and will be evident by the examples that we solve next. Secondly, I may mention that you should think for a moment about how many steps you would have taken to solve example C mentioned above by the traditional method taught at schools. If you were to multiply (9999999 × 9999999) using the traditional system it would have taken many minutes to solve the problem. Hence, undoubtedly the Base Method has a lot of utility in terms of getting instant answers.

Let us solve a few examples:

(Q) Multiply 9750 by 9998.

$$
\begin{array}{r}
10000 \\
9750 - 250 \\
9998 - \quad 2 \\
\hline
9748 / 0500
\end{array}
$$

Since both the numbers are closer to 10000, we take it as the base. The difference between 10000 and 9750 is 250 and the difference between 10000 and 9998 is 2. Next, the base 10000 has four zeros and hence the RHS will be filled by a four-digit answer. Next, we multiply -250 by -2 and write the answer as 0500 (converting it into a four-digit number) and putting it on the RHS. Finally, we subtract 2 from 9750 and put it on the LHS. The final answer is 97480500.

(Q) Multiply 1007 by 1010.

$$
\begin{array}{r}
1000 \\
1007 + \quad 7 \\
1010 + \quad 10 \\
\hline
1017 / 070
\end{array}
$$

In this case, we have the numbers 1007 and 1010. Both the numbers are closer to 1000 and hence we take 1000 as the base. Since, we have both the numbers **above** the base, the difference in this case will be positive and represented by a plus sign. The difference between 1007 and 1000 is (+7) and the difference between 1010 and 1000 is (+10).

Now, since the base 1000 has three zeros, the right hand side will be filled in by a three-digit number. We now multiply the differences 10 by 7 and fill the RHS as 070. Finally, we get the cross answer of 1007 **plus** 10 as 1017 and put it on the RHS. The final answer is 1017070.

Have a look at the following examples:

667 - 333	9988 - 12	9500 - 500	808 - 192
997 - 3	9996 - 4	9991 - 9	999 - 1
664 / 999	9984 / 0048	9491 / 4500	807 / 192

977 - 23	10020 + 20	1230 + 230	123456 + 23456
980 - 20	10020 + 20	1003 + 3	100001 + 1
957 / 460	10040 / 0400	1233 / 690	123457 / 23456

From the examples mentioned above, we can see that this technique works for problems where numbers are above and below the base. But in general usage, we might not always come across such numbers. In some cases we might have numbers like 47, 56, 72, 78 whose difference from the base of 100 is very big. For example, if we were to multiply 47 by 52, the difference would be -53 and -48 respectively and multiplying the differences would itself be a cumbersome task. In some cases, we may have one number above the base and the second number below the base. For example, while multiplying 97 by 103 the difference from the base 100 would be -3 and +3 respectively. In such cases the above mentioned technique will have to be expanded further.

Thus, what we have studied so far is just a small part of the base method. We will now study the various different combinations of numbers and the technique to be used in each case.

(a) When the number of digits in RHS exceeds number of zeros in the base.

We have studied the four steps used in the base method of multiplication. Step B says that the RHS should be filled by as many digits as the number of zeros in the base. In some cases, the answer on the RHS was small and so we filled the remaining places by inserting zeros. An opposite situation may also arise when the number of digits on the RHS is more than the number of zeros in the base. In that case, we will have to carry over as we do in normal multiplication. Have a look at the following examples:

(Q) Multiply 950 by 950.

$$1000$$
$$950 - 50$$
$$950 - 50$$
$$\overline{900 \mid {}_2500}$$

$$= 902500$$

- The base is 1000 and the difference is -50. The number of zeros in 1000 is 3 and so the RHS will be filled in by a three digit answer.
- The vertical multiplication (-50 × -50) gives 2500 and the cross answer gives 900. They are filled in as shown above.
- Note that the RHS can be filled by a three-digit answer only but we have a four-digit number, namely 2500.
- We carry over the extra digit 2 to the LHS and add it to the number 900 and make it 902.
- The number on the LHS is 902 and the number in RHS is 500. The final answer is 902500.

(Q) Multiply 1200 by 1020.

$$1000$$
$$1200 + 200$$
$$1020 + 20$$
$$\overline{1220 \mid {}_4000}$$

$$= 1224000$$

- The base is 1000 and the difference is 200 and 20 respectively.
- The vertical multiplication yields a four-digit answer 4000 which cannot be fitted in three places.
- We carry the extra digit 4 to the LHS and add it to 1220 and make it 1224. The final answer is 1224000.

(Q) Multiply 150 by 140.

$$100$$
$$150 + 50$$
$$\underline{140 + 40}$$
$$190 \mid {}_{20}00$$

$$= 21000$$

- The base is 100 and the difference is 50 and 40 respectively. The number of digits on the RHS will be 2.
- Vertical multiplication is 2000, a four-digit number.
- But the RHS can only be filled in with two digits. We carry over the two excess digits 20 to the LHS and add it to 190. The answer is 210.
- The LHS is 210 and the RHS is 00. The final answer is 21000.

More examples:

100	**1000**	**10000**
$112 + 12$	$1300 + 300$	$9200 - 800$
$\underline{110 + 10}$	$\underline{1020 + 20}$	$\underline{9200 - 800}$
$122 \mid {}_{1}20$	$1320 \mid {}_{6}000$	$8400 \mid {}_{64}0000$
$= 12320$	$= 1326000$	$= 84640000$

10	**1000**	**100**
17 + 7	850 - 150	75 - 25
18 + 8	993 - 7	95 - 5
25 $\mid_5$ 6	843 $\mid_1$ 050	70 $\mid_1$ 25
= 306	= 844050	= 7125

(b) Multiplying a number above the base with a number below the base

We have dealt with examples where both the numbers are below a certain base or above it. We solved examples like 96 by 98 where both the numbers are below the base. We also solved examples like 1003 by 1050 where both the numbers are above the base. Now we will solve examples where one number is above a base and another number is below it.

(Q) Multiply 95 by 115.

$$
\begin{array}{r}
\mathbf{100} \\
95 - 5 \\
\underline{115 + 15} \\
110 \mid (-75)
\end{array}
$$

= 110 × 100 - 75

= 11000 - 75

= 10925

- The base is 100 and the difference is -5 and +15 respectively.
- The vertical multiplication of -5 and +15 gives -75.
- The cross answer is 110.
- At this point, we have the LHS and the RHS. Now, we multiply LHS with the base and subtract the RHS to get the final answer. Thus, 110 multiplied by 100 minus 75 gives 10925.

(Q) Multiply 1044 by 998.

$$\textbf{1000}$$

$$
\begin{array}{r}
1044 + 44 \\
998 - 2 \\
\hline
1042 \quad (-088)
\end{array}
$$

$$= 1042 \times 1000 - 88$$

$$= 1042000 - 88$$

$$= 1041912$$

- The base is 1000 and the difference is 44 and -2 respectively.
- The product of 44 and -2 is 88 which is converted in a three-digit answer and written in RHS.
- The cross answer of 1044 minus 2 is 1042 which is written on the LHS.
- We multiply LHS with the base and subtract RHS from it. Thus, 1042 multiplied by 1000 - 88 gives 1041912.

(Q) Multiply 1,00,032 by 99,990.

$$\textbf{1,00,000}$$

$$
\begin{array}{r}
100032 + 32 \\
99990 - 10 \\
\hline
100022 \mid (-00320)
\end{array}
$$

$$= 100022 \times 100000 - 320$$

$$= 10002199680$$

The base is one lakh and the difference is 32 and -10. The vertical multiplication is -000320. In this case, we multiply the LHS 100022 with the base 1,00,000 and subtract the RHS -320 from it to get the final answer. The final answer is 10002199680.

More examples:

1000	**100**	**10**
800 - 200	120 + 20	14 + 4
1004 + 4	97 - 3	9 - 1
804 │ (-800)	117 │ (-60)	13 │ (-4)
= 804000 - 800	= 11700 - 60	= 130 - 4
=803200	= 11640	= 126

(c) Multiplying numbers with different bases

In the examples solved above, we have multiplied numbers which are closer to the same base. The obvious question that would next arise is how to multiply numbers that have different bases. Suppose we want to multiply 877 by 90. In this case the first number is closer to the base 1000 and the second number is closer to the base 100. Then, how do we solve the problem. Have a look at the examples below:

(Q) Multiply 85 by 995.

Here, the number 85 is close to the base 100 and the number 995 is close to the base 1000. We will multiply 85 with 10 and make both the bases equal thus facilitating the calculation. Since, we have multiplied 85 by 10 we will divide the final answer by 10 to get the accurate answer.

1000

850 - 150
995 - 5
845 │ 750

845750 divided by 10 gives 84575

- We multiply 85 by 10 and make it 850. Now, both the numbers are close to 1000 which we will take as our base.
- The difference is -150 and -5 which gives a product of 750.

- The cross answer is 845 which we will put on the LHS.
- Thus, the complete answer is 845750. But, since we have multiplied 85 by 10 and made it 850 we have to divide the final answer by 10 to get the effect of 85 again. When 845750 is divided by 10 we get 84575.
- Thus, the product of 85 into 9995 is 84575.

(Q) Multiply 102 by 999.

The first number 102 is close to the base 100 and the second number 999 is close to the base 1000. We multiply 102 by 9 and make it 918 which is closer to the base 1000. Thus, we will be multiplying 918 by 999 which is closer to 1000.

$$1000$$

$$
\begin{array}{r}
918 - 82 \\
999 - 1 \\
\hline
917 \mid 082
\end{array}
$$

917082 divided by 9 equals 101898

The answer that we have obtained is 917082. Since, we had multiplied 102 by 9 to get 918 we have to divide the final answer by 9 to get the accurate answer. Thus, 917082 divided by 9 is 101898 which is our final answer.

(Q) Multiply 9995 by 86.

9995 is closer to 10000 and 86 is closer to 100. We multiply 86 with 100 and make it 8600. Thus, both the numbers are closer to 10000.

$$10000$$

$$
\begin{array}{r}
9995 - 5 \\
8600 - 1400 \\
\hline
8595 \mid 7000
\end{array}
$$

The answer that we obtain is 85957000. We divide it by 100 to get the accurate answer. Thus, the product of 9995 by 86 is 859570.

(Q) 73 multiplied by 997.

In this case we multiply 73 by 10 and make it 730. Now both the numbers are closer to the base 1000.

$$1000$$

$$730 - 270$$
$$\underline{997 - 3}$$
$$727 \mid 810$$

= 727810 divided by 10

= 72781

(Q) Multiply 73 by 990.

Here, too, we multiply 73 by 10 and make it 730.

$$1000$$

$$730 - 270$$
$$\underline{990 - 10}$$
$$720 \mid_2 700$$

= 722700

722700 divided by 10 is 72270..

Thus, 73 into 990 is 72270.

In this example, the number of digits on the RHS exceeds the number of zeros in base and hence we carry over the 2 to the LHS.

(Q) Multiply 99 by 1005.

We multiply 99 by 10 and make it 990.

$$1000$$

$$990 - 10$$
$$\underline{1005 + 5}$$
$$995 \mid (-50)$$

= 995000 - 50

= 994950

The answer obtained by dividing 994950 by 10 is 99495. Thus, 99 multiplied by 1005 is 99495.

(d) When the base is not a power of ten

In the last few pages, we have seen the different situations in which the base method was used. We took numbers above and below the base, we took numbers far away from the base and so on. All the while we were taking only powers of 10, namely, 10, 100, 1000, etc. as bases, but in this section we will take numbers like 40, 50, 600, etc. as bases.

In the problems that will follow, we will have two bases — an actual base and a working base. The actual base will be the normal power of ten. The working base will be obtained by dividing or multiplying the actual base by a suitable number. Hence, our actual bases will be numbers like 10, 100, 1000, etc. and our working bases will be numbers like 50 (obtained by dividing 100 by 2) or 30 (obtained by multiplying 10 by 3), 250 (obtained by dividing 1000 by 4).

Actual bases: 10, 100, 1000, etc.

Working bases: 40, 60, 500, 250, etc.

Note: This section requires a great deal of concentration on the part of the reader as the chances of making a mistake are high. Go through the examples carefully.

(Q) Multiply 48 by 48.

Actual base = 100

Working base: 100/2 = 50

$$
\begin{array}{r}
48 - 2 \\
\underline{48 - 2} \\
2)\ \underline{46\ /\ 04} \\
23\ \ /\ \ 04
\end{array}
$$

In this case the actual base is 100 (therefore RHS will be a two-digit answer). Now, since both the numbers are close to 50 we have taken 50 as the working base and all other calculations are done with respect to 50.

The difference of 48 and 50 is -2 and so we have taken the difference as minus 2 in both the multiplicand and the multiplier.

We have the base and the difference. The vertical multiplication of -2 by -2 gives 4. We convert it to a two-digit answer and write it as 04. The cross answer of 48 - 2 is written as 46 and put on the LHS. The answer is 4604.

Now, since 50 (the working base) is obtained by dividing 100 (the actual base) by 2, we divide the LHS 46 by 2 and get 23 as the answer on LHS. The RHS remains the same. The complete answer is 2304.

(In this system, the RHS always remains the same.)

OR

In the example above, we obtained the working base 50 by dividing the actual base 100 by 2. Another option is to multiply the actual base 10 by 5 to make it 50. The second option is explained below:

Actual base: 10

Working base: $10 \times 5 = 50$

$$
\begin{array}{r}
48 - 2 \\
\underline{48 - 2} \\
46 \,/\, 4 \\
\underline{\times \quad 5} \\
230 \,/\, 4
\end{array}
$$

In this case the actual base is 10 (therefore RHS will be a one-digit answer). Now, since both the numbers are close to 50 we have taken 50 as the working base. Since, 50 is obtained by multiplying 10 by 5 we multiply the LHS 46 by 5 and get the answer 230. The RHS remains the same. The complete answer is 2304.

Carefully observe both the cases given above. In the first case we took the actual base as 100 and got a working base 50 on dividing it by 2. In the next case, we got the actual base 10 and got the working base 50 by multiplying it by 5. The student can solve the problem by either system as the answer will be the same.

(Q) Multiply 27 by 28.

Actual base: 100

Working base: $100/5 = 20$

$$
\begin{array}{r}
27 + 7 \\
\underline{28 + 8} \\
35 \;/\; 56 \\
\underline{5)35 \;/\; 56} \\
7 \;/\; 56
\end{array}
$$

The answer obtained through the base method is 3556. But, since 100 is divided by 5 to get 20, we divide the answer on the LHS, namely, 35 by 5 to get 7. The complete answer is 756.

(Q) Multiply 59 by 58.

Actual base: 10

Working base: $10 \times 6 = 60$

$$
\begin{array}{r}
59 - 1 \\
\underline{58 - 2} \\
57 \;/\; 2 \\
\underline{\times\; 6 \;/\; 2} \\
342 \;/\; 2
\end{array}
$$

The actual base is 10 and thus RHS will be a single digit answer. Since the actual base is multiplied by 6 to get the working base, the answer on the LHS is also multiplied by 6 (57×6) to get 342. The complete answer is 3422.

More examples:

(a) 31×31

    (AB – 10, WB -10 × 3 = 30)

$$\begin{array}{r} 31 + 1 \\ \underline{31 + 1} \\ 32 / 1 \\ \times \ \underline{3} \\ 96 / 1 \end{array}$$

(b) 57×57

    (AB – 100, WB – 100/2 = 50)

$$\begin{array}{r} 57 + 7 \\ \underline{57 + 7} \\ 2)\ 64\ /\ 49 \\ 32\ /\ 49 \end{array}$$

(c) 395×396

    (AB -100, WB -100 × 4 = 400)

$$\begin{array}{r} 395 - 5 \\ \underline{396 - 4} \\ 391 / 20 \\ \times \ \ \underline{4} \\ 1564 / 20 \end{array}$$

(d) 228×246

    (AB - 1000, WB - 1000/4 = 250)

$$\begin{array}{r} 228\ -\ 22 \\ \underline{246\ -\ \ 4} \\ 4)\ 224\ /\ 088 \\ 56\ /\ 088 \end{array}$$

(e) 45×45

(AB -10, WB -10 $\times$ 4 = 40)

$$
\begin{array}{r}
45 - 5 \\
45 - 5 \\
\hline
40 / 25 \\
\times \quad 5 \\
\hline
200 /\,_2 5 \\
\hline
= 202/5
\end{array}
$$

(f) 58×42

(AB - 10, WB - 10 $\times$ 5 = 50)

$$
\begin{array}{r}
58 + 8 \\
42 - 8 \\
\hline
50 / (-64) \\
\times \ 5 \\
\hline
250 / (-64) \\
\hline
= 243 / 6
\end{array}
$$

(g) 55×45

(AB - 100, WB -100/2 = 50)

$$
\begin{array}{r}
55 + 5 \\
45 - 5 \\
\hline
2)\ \underline{50 / (-25)} \\
25 / (-25) \\
\hline
= 2475
\end{array}
$$

(h) 45×42 (Important)

(AB - 100, WB - 100/2 = 50)

$$
\begin{array}{r}
45 - 5 \\
42 - 8 \\
\hline
2)\ \underline{37 / 40} \\
18^{1/2} / 40 \\
\hline
= 18/(50 + 40) \\
= 1890
\end{array}
$$

(i) 245×248 (Important)
(AB - 1000, WB - 1000/4)

$$
\begin{array}{r}
245 - 5 \\
\underline{248 - 2} \\
4)\ \underline{243\ /\ 010} \\
60^{3/4}\ /\ 010 \\
= 60\ /\ (750 + 10) \\
= 60760
\end{array}
$$

Have a look at examples (h) and (i). In example (h), the actual base is 100 and the working base is 50. The answer obtained is 3740 and the LHS when divided by 2 gives an answer $18^{1/2}$. However, we cannot have a fraction in the final answer. Hence, we multiply the fraction by the actual base and add it to the RHS. In this case, the fraction $^{1/2}$ is multiplied by the actual base 100 ($^{1/2} \times 100 = 50$) and the answer is added to the RHS. The RHS now becomes 50 plus 40 equal to 90.

In example (i) the actual base is 1000 and the working base is 250. The LHS of the answer, namely, 243 when divided by 4 gives $60^{3/4}$. The fraction is multiplied by the actual base and added to the RHS. Thus, $^{3/4}$ multiplied by 1000 gives 750 which is added to the RHS. The RHS now becomes $10 + 750 = 760$.

With this we complete our study of the Base Method. We have studied all the different cases in which the Base Method can be applied with a sufficient number of solved examples. The last sub-section appears a little confusing at first sight but you will be able to tackle problems after some practice.

Given below is a comparison — the traditional system of multiplication with the Vedic method studied in this chapter.

(Q) Multiply 9996 by 9994.

Traditional Method	**Vedic method**

$$
\begin{array}{r}
9996 \\
\times\ 9994 \\
\hline
39984 \\
889640 \\
8896400 \\
+\ 88964000 \\
\hline
99900024 \\
\end{array}
$$

Vedic method:

10000

9996 - 4

9994 - 6

9990 / 0024

From the comparison it is evident that the Vedic method scores over the traditional method in terms of speed and simplicity by a huge margin.

There are three widely used multiplication systems in the world. The first method is the one taught at schools and we have been using it for our daily calculations (traditional method). The second method is the Criss-Cross system which you have learnt in the earlier part of this book. The third method is the Base Method which you have studied in this chapter. Thus, you are well equipped with three different methods. Whenever you are confronted with a problem on multiplication, you can select a system that you want. If the number is closer to any of the bases or their sub-multiples you can use the Base Method. If it is not close to any such number then you can use the Criss-Cross system.

EXERCISE

Q. (1) Multiply the following numbers:

(a) 990 × 994

(b) 999993 × 999999

(c) 1002 × 10100

(d) 1050 × 1005

Q. (2) Multiply the following numbers where the answer in RHS exceeds the number of zeros in the base.

(a) 16×17

(b) 1500×1040

(c) 9300×9500

(d) 860×997

Q. (3) Calculate the product of the following (one number is above the base and the other number is below the base).

(a) 96×104

(b) 890×1004

(c) 10080×9960

(d) 970×1010

Q. (4) Multiply the following numbers using different bases.

(a) 73×997

(b) 94×990

(c) 82×9995

(d) 102×1010

Q 5) Multiply the numbers using actual and working bases.

(a) 49×48

(b) 22×22

(c) 53×49

(d) 18×17

(e) 499×496

CHAPTER 7

Base Method for Squaring

We have seen the applications of the Base Method in multiplication of numbers. There is a corollary of the Base Method called the 'Yavadunam' rule. This rule is helpful in squaring numbers.

In this chapter we will study the Yavadunam rule and its applications in squaring numbers.

RULE

Swamiji had coined the Yavadunam rule in a Sanskrit line. When translated into English, it means:

'Whatever the extent of its deficiency, lessen it to the same extent and also set up the square of the deficiency.'

We thus see that the rule is composed of two parts. The first part says that whatever the extent of the deficiency, we must lessen it to the same extent. The second part simply says — square the deficiency.

While writing the answer we will put the first part on the LHS and the second part on the RHS.

Let us have a look at an example:

(Q) Find the square of 8.

- We take the nearest power of 10 as our base (In this case 10 itself).
- As 8 is 2 less than 10, we should decrease 8 further by 2 and write the answer so obtained, viz. 6 as the left hand part of the answer.
- Next, as the rule says, we square the deficiency and put it on the RHS. The square of 2 is 4 and hence we put it on the RHS.
- The LHS is 6 and RHS is 4. The complete answer is 64. Thus, the square of 8 is 64.

$$10$$
$$8 - 2$$
$$\underline{8 - 2}$$
$$6 / 4$$

Similarly, the square of 9 is 81.

$$10$$
$$9 - 1$$
$$\underline{9 - 1}$$
$$8 / 1$$

Since the Yavadunam rule is a corollary of the Base Method, the method used in this rule is exactly similar to the Base Method. However, we represent it in a different manner. Look at the examples given below:

(Q) Find the square of 96.

- In this case, we take the nearest power of ten, viz. 100.
- The difference of 100 and 96 is 4 and so we further subtract 4 from 96 and make it 92.
- We square 4 and make it 16 and put it on the RHS.
- The complete answer is 9216.

(Q) Find the square of 988.

- We take the nearest power of 10, i.e. 1000 as base.
- The difference of 988 and 1000 is 12 and therefore we subtract 12 from 988 and make it 976. This becomes the left half of our answer.
- We square twelve and put it on the RHS as 144.
- Thus, the square of 988 is 976144.

(Q) Find the square of 97.

- We take the nearest power of 10 namely 100 as base.
- The difference between 100 and 97 is 3 and therefore we further remove 3 from 97 and make it 94.
- Now, we square 3 and write it as 09 (because the base has two zeros) and put it on the RHS.
- The complete answer is 9409.

We have seen how the Yavadunam rule can be used in squaring numbers that are below a certain power of ten. In the same way, we can also use the rule to square numbers that are above a certain power of ten. However, instead of decreasing the number still further by the deficit we will increase the number still further by the surplus.

(Q) Find the square of 12.

- We take the nearest power of 10 closer to 12 which is 10 itself.
- The difference between 10 and 12 is 2 and so we further add 2 to 12 and make it 14. This becomes our LHS.
- We square the surplus 2 and make it 4 and this becomes our RHS.
- Thus, the square of 12 is 144.

(Q) Find the square of 108.

- We take 100 as the base and 8 as the surplus.

- We further add the surplus 8 to the number and make it 116.

- We square the surplus and make it 64.

- The final answer is 11664.

(Q) Find the square of 14.

- We take 10 as the base and 4 as the surplus.

- We further add 4 to 14 and make it 18 (LHS).

- We square 4 and write 16 as the RHS.
  $14^2 = 18/_16 = 19/6$

(We have observed in the previous chapter that if the base is 10, the RHS can be a single digit answer only. In this case, it is a two-digit answer, namely, 16. Hence, we carry over the extra digit 1 to LHS and add it to 18).

(Q) Find the square of 201.

Before we use the Yavadunam rule to solve the question, let us recall how the Base Method would have been used to solve it.

Base Method:

$$\text{Actual Base: } 100$$
$$\text{Working Base: } 100 \times 2 = 200$$

$$
\begin{array}{r}
201 + 1 \\
\underline{201 + 1} \\
202 \,/\, 01 \\
\times \quad 2 \\
\hline
404 \,/\, 01
\end{array}
$$

We multiplied the actual base by 2 to get the working base. Thus, we multiplied the LHS by 2 to get the final answer.

Using the Yavadunam rule:

- The actual base is 100 and the working base is 200 (100 × 2).

- The surplus of 201 over 200 is 1 and so we further increase it by 1 making it 202. This becomes our LHS.
- We square the surplus 1 and put it on the RHS after converting it into a two-digit number, viz. 01.
- The complete answer is 202/01.
- But since we have multiplied the actual base by 2 to get the working base, we multiply the LHS by 2 and make it 404. The RHS remains the same.
- The final answer is 404/01.

More Examples:

(a) $93^2 = 86/49$

(b) $997^2 = 994/009$

(c) $13^2 = 16/9$

(d) $15^2 = 20/_2 5 = 22/5$

(e) $960^2 = 920/_1 600 = 921/600$

(f) $9985^2 = 99700225$

(g) $202^2 = 40804$

(h) $301^2 = 90601$

EXERCISE

Find the squares of the following numbers using the Yavadunam Rule.

PART A

    (1) 7

    (2) 95

    (3) 986

    (4) 1025

    (5) 1012

PART B

    (1) 85

    (2) 880

(3) 910
(4) 18
(5) 1120

PART C

(1) 22
(2) 203
(3) 303
(4) 498 (Hint: Take working base as 1000/2 = 500)
(5) 225 (Hint: Take working base as 1000/4 = 250)

CHAPTER **8**

Digit-Sum Method

All the techniques that we have discussed till now have emphasized various methods of quick calculation. They have helped us in reducing our time and labour and in some cases provided the final answer without any actual calculation.

In this chapter, we will study the digit-sum method. This method is not used for quick calculation but for quick checking of answers. It will help us verify the answer that we have obtained to a particular question. This technique has wonderful different applications for students giving competitive exams as they are already provided with four alternatives to every answer.

Although the digit-sum method is discussed by Jagadguru Bharati Krishna Maharaj in his study, mathematicians in other parts of the world were aware of this principle even before the thesis of Swamiji was published. Prof. Jackaw Trachtenberg and other mathematicians have dealt with this principle in their research work.

METHOD

The digit-sum method (also known as the add-up method) can

be used to check answers involving multiplication, division, addition, subtraction, squares, square roots, cube roots, etc. The procedure involved is very simple. We have to convert any given number into a single digit by repetitively adding up all the digits of that number.

Example: Find the digit-sum of 2467539

Ans: The number is 2467539.
       We add all the digits of that number.
       $2 + 4 + 6 + 7 + 5 + 3 + 9 = 36$

Now, we take the number 36 and add its digits $3 + 6$ to get the answer 9.

Thus, we have converted the number 2467539 into its digit-sum 9.

Example 2: Find the digit-sum of 56768439

Ans: $5 + 6 + 7 + 6 + 8 + 4 + 3 + 9 = 48$
       $4 + 8 \qquad\qquad\qquad = 12$
       $1 + 2 \qquad\qquad\qquad = 3$

Hence, the digit-sum of 56768439 is 3.

(Note: the digit sum will always be a single digit. You have to keep on adding the numbers until you get a single-digit answer.)

A few more examples are given below:

NUMBER	PROCESS	DIGIT-SUM
34	$3 + 4 = 7$	7
4444	$4+4+4+4 = 16$ & $1+6 = 7$	7
915372468	$9+1+5+3+7+2+4+6+8 = 45$ & $4+5 = 9$	9
47368	$4+7+3+6+8 = 28$ & $2+8 = 10$ & $1+0 = 1$	1

We have discussed how to calculate the digit-sum of a number. We shall now solve a variety of illustrated examples involving different arithmetical operations.

Example 1 (Multiplication)

(Q) Verify whether 467532 multiplied by 107777 equals 50389196364.

Ans: First we will calculate the digit-sum of the multiplicand. Then we will calculate the digit-sum of the multiplier. We will multiply the two digit-sums thus obtained. If the final answer equals to the digit sum of the product then our answer can be concluded to be correct.

The digit sum of 467532 is 9.

The digit sum of 107777 is 2.

When we multiply 9 by 2 we get the answer 18. Again the digit sum of 18 is 9.

Thus, the digit-sum of the completed multiplication procedure is 9.

Now, we will check the digit-sum of the product.

The digit sum of 50389196364 is also 9. The digit sum of the question equals to the digit sum of the answer and hence we can assume that the product is correct.

Example 2 (Division)

(Q) Verify whether 2308682040 divided by 36524 equals 63210.

Ans: We can use the formula that we had learnt in school.

Dividend = Divisor x Quotient + Remainder.

In this case we will be using the same formula but instead of the actual answers we will be using their digit-sums.

The digit-sum of Dividend is 6.

The digit sum of divisor, quotient and remainder is 2, 3, and 0 respectively.

Since $6 = 2 \times 3 + 0$, we can assume our answer to be correct.

In this manner, we can solve sums involving other

operations too. However, before continuing ahead I will introduce a further short-cut to this method. The rule says:

While calculating the digit-sum of a number, you can eliminate all the nines and all the digits that add up to nine.

When you eliminate all the nines and all the digits that add up to nine you will be able to calculate the digit-sum of any number much faster. The elimination will have no effect on the final result.

Let us take an example:

(Q). Find the digit-sum of 637281995.

Ans: The digit sum of 6372819923 is:
6+3+7+2+8+1+9+9+2+3 = 50 and again 5+0 is 5.

Now, we will eliminate the numbers that add up to 9 (6 and 3, 7 and 2, 8 and 1 and also eliminate the two 9's).

We are left with the digits 2 and 3 which also add up to 5. Hence, it is proved that we can use the short-cut method for calculating the digit-sum. The answer will be the same in either case.

A few more examples:

(a)	(b)	(c)	(d)
Number	Normal Digit Sum	Numbers left After Elimination of 9's and digits adding up to 9	Digit Sum After Elimination
199999	1	1	1
63727	7	7	7
45231	6	231	6
8001573	6	573	6
56789	8	5678	8

From the above table we can see that the values in column 'b' and the values in column 'd' are similar.

Note: If the digit-sum of a number is 9, then we can eliminate the 9 straight away and the digit-sum becomes 'zero.'

Example 3 (Multiplication)

(Q). Verify whether 999816 multiplied by 727235 is 727101188760.

Ans: The digit sum of 999816 can be instantly found out by eliminating the three 9's and the combination of 8 plus 1. The remaining digit is 6 (which becomes our digit-sum).

The digit sum of 727235 can be instantly calculated by eliminating the numbers that add up to nine. The digit sum of the remaining digits is 8.

When 8 is multiplied by 6 the answer is 48 and the digit sum of 48 is 3.

But, the digit sum of 727101188750 is 2. The digit-sum of the question does not match with the digit-sum of the answer and hence the answer is certainly wrong.

Example 4 (Addition)

(Q). Verify whether 18273645 plus 9988888 plus 6300852 plus 11111111 is 45674496.

Ans: The digit-sum of the numbers is 0, 4, 6 and 8 respectively. The total of these four digit sum is 18 and the digit sum of 18 is 9.

The digit sum of 45674496 is also 9 and hence the sum is correct.

I think four examples will suffice. In the similar manner, we can check the answer obtained by other mathematical operations too. I have given below a list of the mathematical procedure involved in a particular problem and the technique for calculating the digit-sum.

MATHEMATICAL OPERATION	PROCEDURE FOR CHECKING ANSWER
Multiplication	The digit-sum of the multiplicand when multiplied with the digit-sum of the multiplier should equal to the digit-sum of the product.
Division	Use the formula dividend = divisor multiplied by quotient + remainder. (use digit-sums instead of actual numbers).
Addition	The digit-sum of the final sum should be equal to the digit-sum of all the numbers used in the addition process.
Subtraction	The digit-sum of the smaller number as subtracted from the digit-sum of the bigger number should equal the digit-sum of the difference.
Squaring/Square-Rooting	The digit-sum of the square-root as multiplied by itself should equal to the digit sum of the square. E.g.: To verify whether 23 is the square root of 529. Method: The digit sum of 23 is 5 and when 5 is multiplied by itself the answer is 25. The digit sum of 25 is 7 and the digit sum of 529 is also 7.
Cubing/Cube Rooting	The digit-sum cube-root when multiplied by itself and once again by itself should equal to the digit sum of the cube.

MATHEMATICAL OPERATION	PROCEDURE FOR CHECKING ANSWER
	E.g.: Verify whether 2197 is the cube of 13. Ans: The digit-sum of 13 is 4. 4 multiplied by itself and again by itself is 64. The digit-sum of 64 is 1 and the digit-sum of 2197 is also 1.

APPLICATIONS

The digit-sum method has immense utility for practitioners of Numerology and other occult sciences. The knowledge that they can eliminate the 9's and numbers that add up to 9 makes their task simpler.

For students giving competitive and other exams, this technique has a lot of utility. Many times they can check the digit-sum of each of the alternatives with the digit-sum of the question and try to arrive at the correct answer. This will eliminate the need for going through the whole calculation.

However, there is one drawback with this technique. The drawback is that the digit-sum method can tell us only whether an answer is wrong or not. It cannot tell us with complete accuracy whether an answer is correct or not.

This sentence is so important that I would like to repeat it again.

The digit-sum method can only tell us whether an answer is wrong or not. It cannot tell us with complete accuracy whether an answer is correct or not.

Let me illustrate this with an example.

(Q) What is the product of 9993 multiplied by 9997.

Method: Assume that you have read the question and

calculated the answer as 99900021. The digit sum of the question is 3 and the digit-sum of the answer is 3 and hence we can assume that the answer is correct.

However, instead of 99900021 had your answer been 99900012 then too the digit-sum would have matched even though the answer is not correct. Or for that matter if your answer would have been 99990021 then too the digit-sum would have matched although this answer is incorrect too. Or in an extreme case, even if your answer would have been 888111021 then still the digit-sum would have matched although it is highly deviated from the correct answer!

Thus, even though the digit-sum of the answer matches with that of the question, you cannot be 100% sure of its accuracy. You can be reasonably sure of its accuracy but cannot swear by it.

However, **if the digit-sum of the answer does not match with the digit sum of the question then you can be 100% sure that the answer is wrong.**

In a nutshell,

If the digit-sums	Then the answer is
Match	Most likely correct
Do not match	Wrong

For practitioners of numerology and other occult sciences, there is no question of checking answers and hence this technique can *per se* come to their aid.

EXERCISE

(Note: Before attempting to solve the questions, remember that 9 is synonymous with 0 and can be used interchangeably).

PART A

Q. (1) Instantly calculate the digit-sum of the following numbers:

(1) 23456789

(2) 27690815

(3) 7543742

(4) 918273645

PART B

Q. (1) Verify whether the following answers are correct or incorrect without actual calculation.

(1) $95123 \times 66666 = 6341469918$

(2) 838102050 divided by $12345 = 67890$

(3) $88^2 = 7444$

(4) $88^3 = 681472$

(5) $475210 + 936111 + 315270 = 726591$

(6) $9999999 - 6582170 = 3417829$

(7) $6582170 - 9999999 = -3417829$

(8) $\dfrac{900}{120}$ gives quotient 7 and remainder 60

(9) $0.45632 \times 0.65432 = 0.2985793024$

PART C

Q. (1) Select the correct answer from amongst the alternatives without doing actual calculation (Calculate the digit-sum of each alternative and match it with the question).

(1) $3569 \times 7129 =$ _____

    (1) 25443701

    (2) 25443421

    (3) 25443401

    (4) 25445401

(2) $6524 + 3091 + 8254 + 6324 + 7243 + 5111 + 9902 + 3507 =$ _____

    (a) 49952

    (b) 49852

    (c) 59956

    (d) 49956

Magic Squares

'Magic Squares' is a term given to squares which are filled with consecutive integers and the total of whose rows, columns and diagonals is always the same. When the numbers in any row, column or diagonal are added up they reveal the same total. Many people have found these squares fascinating and so they have been regarded as magic.

This technique is not a part of formal Vedic Mathematics as discovered by Swamiji. However, the oriental schools of astrology, Feng Shui, numerology and other mystical sciences were aware of this technique. Many practitioners of these sciences in ancient India and other parts of the world often used magic squares for their aid. The Vedic seer undoubtedly used the principles of magic squares for various application. I have included this technique in this book because in lower level competitive exams, questions on magic squares are often asked.

A sample magic square is given below. It is a three-by-three grid and you will find that the total of all rows, columns and diagonals is 15. Since there are 9 squares in the grid, we have used numbers from 1 to 9.

4	3	8
9	5	1
2	7	6

We can verify the various totals:

Row 1 : 4 + 3 + 8 equals 15
Row 2 : 9 + 5 + 1 equals 15
Row 3 : 2 + 7 + 6 equals 15

Column 1 : 4 + 9 + 2 equals 15
Column 2 : 3 + 5 + 7 equals 15
Column 3 : 8 + 1 + 6 equals 15

Diagonal 1 : 4 + 5 + 6 equals 15
Diagonal 2 : 2 + 5 + 8 equals 15

Let us have a look at another magic square. This is a five-by-five grid and the total of all sides will add up to 65. Since, there are 25 squares in the grid, we have used numbers from 1 to 25.

11	10	4	23	17
18	12	6	5	24
25	19	13	7	1
2	21	20	14	8
9	3	22	16	15

You may verify the total of the rows, columns and diagonals. They will all add up to 65.

The obvious question is, How are these magic squares formed? In my high school years, I used to show sample magic

squares to my friends and tell them that I had formed these
squares out of a lot of calculations and taxing my brain for
many hours! However, the truth is that these squares can be
formed very easily and once you understand the concepts
behind them you can form squares of any size. One of my
students who is just nine-years of age formed a magic square
with 21 rows and 21 columns within ten minutes!

You will soon see for yourself that forming these squares is
child's play. I have explained the rules of forming such squares
below. I urge you to read the rules carefully 3-4 times.

RULES:

(1) Always put the number 1 in the centre-most square of
the last column.
(2) After inserting a number in a square move to the square
in the south-east direction and fill it with the next number.
(3) If the square in the south-east direction cannot be filled,
then move to the square in the west and fill it with the
next number.
(4) When you have filled a number in the last square of the
grid, fill the next number in the square to its west.

These are the four rules that we will be following. However,
they will not be followed exactly as mentioned above. We will
apply them in a slightly different manner.

To make the magic square, we shall be making some
imaginary squares to effectively follow the four rules mentioned
above. But, I shall not explain what are these imaginary squares
at this point of time. We shall deal with them straight away in
the illustrative example given below:

For a thorough and in-depth understanding of the four rules,
we will take the example of a five by five grid and make a
magic square out of it. The numbers used will be from 1 to 25.

A key of directions is given below for your convenience:

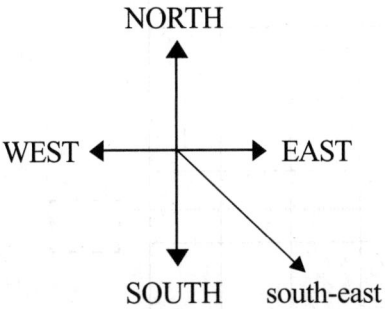

(a) First, we follow rule 1 and place the first number 1 in the centre-most square of the last column

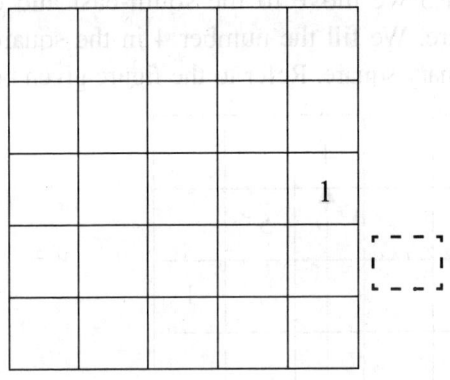

(b) Next, we move to the south-east direction from this square. However, there is nothing in the south-east direction and hence we have to create an imaginary square in the south east direction. As per the rules the digit 2 will come in the imaginary square. However, you cannot put anything in an imaginary square and hence the number 2 will be written in a square farthest from it. Hence, the number 2 will come as shown below.

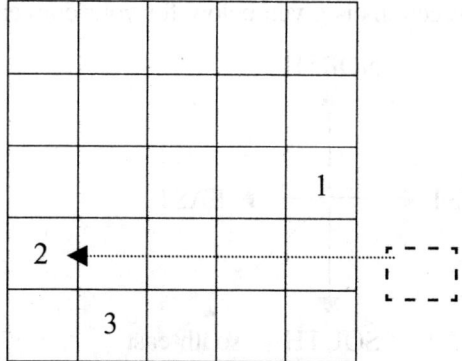

(c) Now from the square where we have written the number 2 we come to the south-east and fill the number 3.

(d) Next, from 3 we move to the south-east and create an imaginary square. We fill the number 4 in the square farthest from this imaginary square. Refer to the figure given below:

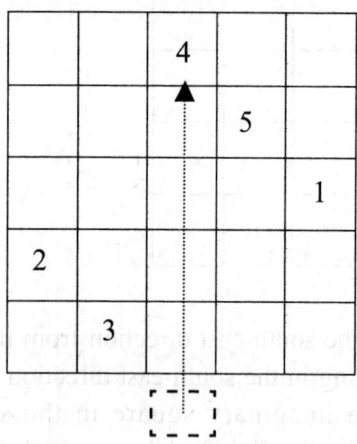

(e) From 4 we come to the south-east and fill the number 5 in the square to its south-east.

(f) Next, from 5 we have to come to the south-east to fill the number 6. But, we realize that the square in the south-east of 5 is already occupied by the digit 1. In this case we will follow the Rule 3 as mentioned above which reads:

If the square in the south-east direction cannot be filled, then move to the square in the west and fill it with the next number

Hence, we will fill the number 6 in the square which lies to the 'west' of 5. Refer to the diagram given below:

		4		
		6	5	
			7	1
2				8
	3			

(g) Next, from 6 we move to the south-east and write the number 7. From 7, we move to the south-east and fill the number 8.

(h) From 8, we move to the south east and create an imaginary square. We fill the number 9 in the square farthest from this imaginary square in the same direction. (Refer to the figure given below.)

(i) From 9, we go to the south-east direction and make an imaginary square and fill the number 10 in the square farthest from this imaginary square in the same direction.

11	10	4		
	12	6	5	
		13	7	1
2			14	8
9	3			15

(j) From 10, we move towards the south-east and find that the square is already occupied by the number 6. Hence, we move to the west of 10 and fill it with the number 11.

(k) From 11 we move to the south-east and fill 12. From 12 we move to the south-east and fill 13 and from 13 we move to 14 and from 14 we move to 15.

(l) We have written the number 15 in the last square of the grid. Hence, we will have to use Rule (4) over here which reads:

'When you have filled a number in the last square of the grid, fill the next number in the square to its west.'

Thus, we will fill the number 16 in the square to the 'west' of 15.

11	10	4		
	12	6	5	
		13	7	1
2			14	8
9	3		16	5

(m) From 16, we create an imaginary square and fill 17. From 17 we create an imaginary square and fill 18. From 18 we come to the south-east and fill 19. From 19, we come to the south-east and fill 20.

11	10	4		17
18	12	6	5	
	19	13	7	1
2		20	14	8
9	3		16	15

(n) From 20, we cannot go to the south-east and have to come to the west to fill the number 21. From 21 we move to the south-east and fill the number 22. From 22, we create an imaginary square and fill 23 in the square farthest from it in the same direction. From 23, we come to the south-east and fill 24. From 24, we create an imaginary square and fill the last number 25.

11	10	4	23	17
18	12	6	5	24
25	19	13	7	1
2	21	20	14	8
9	3	22	16	15

With this, we have just completed our first magic square!

All the numbers from 1 to 25 are fitted into their proper places in the grid. You will find that the total of each row, column and grid equals 65.

I sincerely request the reader to take a blank page and make a five-by-five grid and try to make a magic square exactly as shown above without referring to the procedure that I have explained. If you get blank at any point of time, you may refer to the explanation that I have provided.

Also, I want you to refer to the magic square of the three-by-three grid given at the beginning of this chapter and verify whether it follows the rules that we have learnt.

After you have done both the activities mentioned above, I would like to draw your attention to some of the properties of magic squares. Understanding these properties will help you effectively deal with problems on magic squares.

THE PROPERTIES OF MAGIC SQUARES

(a) The number of rows and columns will always be equal and it will always be an odd number. You can form a three-by-three grid, five-by-five grid, seven-by-seven grid, but using this system you cannot form a magic square of a two-by-two grid, four-by-four grid, six-by-six grid, etc.

(b) The first and last numbers always lie in the same row and exactly opposite to each other.

(c) The total of any side can be found out by multiplying the number in the centre-most square of the grid with the number of squares in any side. For example, in a three by three grid, the centre-most number is 5 and the number of squares on any one side is 3. When 5 is multiplied by 3 the answer is 15 and hence the total of any side is always 15. In the example of a five-by-five grid, the centre-most number is 13 and the number of squares in any side is 5 and hence the total of any row, column or diagonal will always be 65.

(d) You can find out which number will come in the centre-most square of any grid by

Taking the maximum number, dividing it by 2 and rounding it off to the next higher number

Example: In a three-by-three grid, the maximum number is 9. When 9 is divided by 2 the answer is 4.5 and when 4.5 is rounded off to the next higher number the answer is 5. Hence, the centre-most square in a three-by-three grid will be occupied by the number 5.

Example 2: A five-by-five grid will have 25 squares and so the maximum number will be 25. When 25 is divided by 2 the answer is 12.5 and when 12.5 is rounded off the answer is 13. Hence, the number that will appear in the centre-most square of a five-by-five grid will be 13.

(e) Combining properties (c) and (d), we can find the total of a side of any magic square. Suppose you want to find out what will be the total of any side of a seven-by-seven grid.

We know that there are 49 squares in a seven-by-seven grid and hence the highest number is 49. 49 divided by 2 and rounded off gives 25. Hence, the centre-most square will be occupied by the number 25. Next we want to find the total of any side. We multiply the number in the centre-most square by the number of squares in any side. In this case, the answer will be 25 multiplied by 7 which is 175. Hence, the total of any side of a seven-by-seven grid will be 175.

On similar lines, it can be proved that the total of any side of a nine-by-nine grid will be 81 divided by 2 and rounded off which is 41 and then 41 multiplied by 9 which is 369.

(f) There are many possible ways by which a magic square can be made out of a certain grid. Let us suppose we take a three-by-three grid, then we can form a magic square out of it in a few different ways as given below:

4	3	8
9	5	1
2	7	6

8	1	6
3	5	7
4	9	2

6	7	2
1	5	9
8	3	4

2	9	4
7	5	3
6	1	8

These are few of the many possible ways of making a magic square out of a certain type of grid. To understand the logic that was used in this case, we will have to refer to rule 1 mentioned above. Rule 1 says that we must fill the number 1 in the centre-most square of the last column. However, we can tilt the square in any of the four directions and make any column as our last column and thus we will have four possible answers. Observe the number 1 in each of the four cases given above and you will realize that the square has been tilted in four different directions and hence the number 1 appears in the centre-most square of the last column in each case.

Since we have understood the properties of magic squares and also learnt how to make them, the intention of this chapter is fulfilled. I have given below a nine-by-nine grid which I want the reader to study carefully and understand how the numbers are placed one after the other.

The grid will be followed with a practice exercise:

37	36	26	16	6	77	67	57	47
48	38	28	27	17	7	78	68	58
59	49	39	29	19	18	8	79	69
70	60	50	40	30	20	10	9	80
81	71	61	51	41	31	21	11	1
2	73	72	62	52	42	32	22	12
13	3	74	64	63	53	43	33	23
24	14	4	75	65	55	54	44	34
35	25	15	5	76	66	56	46	45

EXERCISE

PART A

(a) Make a three by three grid using the first 9 even numbers (2,4,6....18).

(b) Will the total of each side of such a grid (as created using question 'a' given above) be equal?

(c) What will be the number in the centre-most square of such a grid?

(d) What will be the total of each side of this grid?

PART B

(a) Make a five-by-five grid using the multiples of 3 (3,6,9, etc.)

(b) If you start the grid mentioned above with the number 15 (instead of the number 3) will the sides still give equal total?

(c) What will be the difference in the value of the centre-most square as computed in case 'a' and case 'b' above?

(d) What will be the difference in the total as computed in case 'a' and case 'b' mentioned above?

PART C

(a) Using the multiples of 5, make a three-by-three grid and represent it in four different ways.

(b) A man has 49 cows. The first cow gives one litre milk, the second cow gives two litres of milk, the third cow gives three litres of milk and so on upto the forty-ninth cow which gives 49 litres of milk. This man has seven sons. He wants to divide the 49 cows with his seven sons in such a manner that each child receives the same number of cows and the same quantity of milk. Can you help him do it?

Dates & Calendars

A book integrating the syllabus of various academic and competitive exams cannot be complete without the mention of techniques dealing with dates and calendars. In the recent past, many competitive exams have asked questions relating to dates and calendars.

In this chapter we shall study certain characteristics of dates and the corresponding days on which they fall. We will also discuss a technique which will help you find the day on which any date falls for the entire century!

In my book *How To Top Exams & Enjoy Studies,* I had mentioned a technique by which one can find out the day on which any date falls in the current year. Many readers sent me e-mails appreciating the concept and urged me to provide more information on the topic. Thus, I have included such techniques in this book.

I have given below a technique to find the day on which any date falls for a single year. This technique can work for one year only.

SINGLE YEAR CALENDAR

Let us assume we are dealing with the year 2012.

Given below is the key by which you will be able to tell me the day on which any date falls for the year 2012.

Key

```
    154     163     152     742
```

There are 12 numbers in the box. Each number represents a month of the year. The number 1 represents January, the next number 5 represents February, the next number 4 represents March and so on up to the last number 2 which represents December.

Basically, each of those numbers represents the first Sunday of its corresponding month.

Hence,   1 January is a Sunday
         5 February is a Sunday
         4 March is a Sunday
         1 April is a Sunday
         And so on…

Examples

(i)   What day is 3rd January?
      According to the key, 1st January is a Sunday. Hence, 3rd January is a Tuesday.

(ii) What is the day on 4th August?
      Since August is the eighth month of the year, we use the eighth digit in the key. The eighth digit that

represents August is 5. We know it represents the first Sunday of August. Since 5th August is Sunday and so 4th August is a Saturday.

(iii)   What is the day on 22nd October?

Since October is the tenth month in the year, we look at the tenth digit in the key which is 7. We know that 7th October is a Sunday. Hence the next Sundays will be 14th and 21st October (obtained by adding 7). Since, 21st October is a Sunday and hence 22nd October is a Monday.

This key will work only for the year 2004. Since the first Sundays of every month keep on changing, we will have a different key for every year. The key for the next few years is as given below:

	Jan	Feb	Mar	Apr	May	Jun	Jul	Aug	Sep	Oct	Nov	Dec
2013	6	3	3	7	5	2	7	4	1	6	3	1
2014	5	2	2	6	4	1	6	3	7	5	2	7
2015	4	1	1	5	3	7	5	2	6	4	1	6
2016	3	7	6	3	1	5	3	7	4	2	6	4
2017	1	5	5	2	7	4	2	6	3	1	5	3
2018	7	4	4	1	6	3	1	5	2	7	4	2
2019	6	3	3	7	5	2	7	4	1	6	3	1
2020	5	2	1	5	3	7	5	2	6	4	1	6

Most of my students find this system very helpful. At the commencement of a new year, they take a calendar and make a list of the first Sundays of the twelve months and memorize it. Let us say that the new year 2013 has commenced. On the very first day of the year, they remember the key corresponding to the year 2013. The key is 633 752 741 631 (refer to the table given above). Whenever the need to find out any day/date arises, they recall the key from memory and get the answer. There is no need for an actual calendar.

Now we will study a system for finding the day on which

any date falls for the entire century. After using the technique you will be able to find the day on which any date falls from 1st January 1901 to 31st December 2000.

You may use this technique on your friends, colleagues and relatives. Ask them to tell you the date on which they are born. (5th February 1984 or 15th November 1970, etc.) And on the basis of their date of birth you will be able to instantly tell the day on which they are born!

(Note: The key given below is for the century 1901 to 2000. For the current century, i.e. from 2001 to 2100 the key will be 733614625735. Rest all calculations remain same)

TECHNIQUE

It should be noted here that the techniques mentioned in this chapter are not my invention. Many of these constitute the common wisdom of our profession. Different experts have different ways of predicting the day on which a particular date falls. After studying many such systems and applying them in practice, I found a particular system extremely simple. It involves the use of addition and division only.

Before proceeding with the technique, I will give you a key of the months. This key will remain the same for any given year from 1901 to 2000.

Jan	Feb	Mar	Apr	May	Jun	Jul	Aug	Sep	Oct	Nov	Dec
1	4	4	0	2	5	0	3	6	1	4	6

You will have to memorize this months key. To facilitate memorizing the key, I have made a small verse:

It's the square of twelve
and the square of five
and the square of six
and one-four-six

(Note: The square of 12 is **144**, the square of 5 is **025**, the square of 6 is **036** and then you have **146**.)

The verse will help you to memorize the key easily. I request you to be conversant with this key before proceeding ahead.

In this method of calculation, the final answer is obtained in the form of a remainder. On the basis of this remainder we are able to predict the day of the week.

Remainder	Day
1	Sun
2	Mon
3	Tue
4	Wed
5	Thu
6	Fri
0	Sat

The day key is very simple and needs no learning. Next, we will do an overview of the steps used in predicting the day.

STEPS

- Take the last two digits of the year (If the year is 1942 then take 42)
- Add the number of leap years from 1901
- Add the month-key
- Add the date
- Divide the total by 7
- Take the remainder and verify it with the day-key

These are the six steps that are required to predict the day. Note that the second step requires us to calculate the number of leap years from 1901. A year is a leap year if the last two digits are divisible by 4. Hence, the years 1904, 1908, 1912, 1916, 1920, etc. are all leap years.

If you are asked to calculate any date in the year 1917, then you will have to add the digit 4 in the second step as there are 4 leap years from 1901 to 1917 (1904, 1908, 1912, 1916).

If you are asked to calculate a date in the year 1925, then you will have to add the digit 6 in the second step as there are six leap years from 1901 to 1925.

EXAMPLES

(A) What day is 1 January 1941?
- First, we take the last two digits of the year   = 41
- We add to it the number of leap year from
  1901 to 1941   = 10
- We add the month key for January (refer to
  the month-key)   = 1
- We add the date (as given in the question)   = 1

TOTAL   **53**

When we divide 53 by 7, we get the quotient as 7 and the remainder as 4. However, we are concerned with the remainder only. From the day key, it can be seen that the remainder 4 corresponds with a Wednesday. Hence, 1st January 1941 was a **Wednesday**.

(B) What day is 26th June 1983
- The year will be taken as 83
- Number of leap years (from 1901 to 1980) will be 20
- The month key for June is 5
- The date is 26
- The final total is 134, which gives a remainder of 1 when divided by 7
- The remainder 1 corresponds to a Sunday and hence 26th June, 1983 was a Sunday.

(C) What is the day on 15th August 1947

- Year: 47
- Leap Years: 11
- Month Key: 3
- Date: 15
- (Total 76) when divided by 7 gives a remainder of 6
- Hence, 15th August 1947 was a Friday

Since I was explaining the concept for the first time, I solved example A in detail. However, as me moved to examples B and C we solved them much faster. With enough practice you will be able to predict the day on which a date falls in less than 12-15 seconds. Once the month key and the day key are properly memorized then the remaining calculation is just elementary mathematics.

Another thing to be kept in mind is that when a given year is itself a leap year then it has to be accounted for.

Let us assume that you are asked to find a date in the year 1940. In this case, you will take the last two digits of 1940 and take them as the year in step one. Next, you will calculate the number of leap years from 1901 to 1940. In this case, you have to include the year 1940 as a leap year.

Hence, in step B the number of leap years from 1901 to 1940 will be taken as 10.

If you are asked to find a date in the year 1980, you will take 20 as the number of leap years from 1901 to 1980.

(D) Find the day on 1st July 1944

- Year is 44
- Leap years is 11 (1944 included)
- Month Key is 0
- Date is 1
- Total 55 divided by 7 gives remainder of 6
- Hence, the day is FRIDAY.

As mentioned earlier, this technique works for any given date from 1901 to 2000. However, there is one exception to its working. It is given below:

If the year is a leap year and the month is either January or February, then you have to go one day back from the final answer to arrive at the actual answer.

Thus, we observe that there are two conditions required for a case to be an exception. First, it should be a leap year and second the month has to be January or February.

The exception can be handled by taking an additional step, viz. of going a day backwards from the final answer.

Example: Find the day on 10 February 1948

In this case, we observe that the year is a leap year and the month is February. Hence, it is an exception case.

- First, we take the year as 48
- Next, we add the number of leap years as 12
- The month key for February is 4
- The date is 10
- The total 74 divided by 7 gives a remainder of 4
- According to the day key, a remainder of 4 implies Wednesday
- However, since it is an exception case, we will go a day backwards and get our correct answer as Tuesday
- Hence, 10th February 1948 is a Tuesday

I suggest you try a few dates to test whether you have understood the concepts or not. You can take dates given in the exercise of this chapter and verify it with the solution given at the end of the book.

Next, we will observe certain properties and characteristics of dates. The knowledge of these properties will be of immense utility to students who have to crack date-based problems within time constraints.

CHARACTERISTICS OF DATES

RULE A (applies to all months of the year)

There are exactly 52.1 weeks in a year. In other words, a year is made up of 52 complete weeks and an extra day. Because of this extra day, as every year progresses, a date moves a day later. If May 4 is a Sunday this year, then it will become a Monday in the next year and it will become a Tuesday after two years.

(Note that rule B, mentioned below, applies to the months from March to December only. We have a separate rule C for January and February.)

RULE B (applies to months from March to December)

We know that as every year passes, a date moves a day later. Hence, if it is Thursday this year, then it will become Friday next year. However, when a leap year occurs then a date moves two days later. Thus, if it is a Thursday this year and the next year is a leap year, then it will move two steps ahead and become a Saturday. This phenomenon takes place because the extra day of the leap year pushes the date a further step ahead. Hence, instead of taking a regular one-step ahead, it takes two steps.

Example:

8 July 2005 is a Friday. In 2006, it will move a step ahead and occur on a Saturday. In 2007, it will occur on a Sunday. In 2008, (which is a leap year) it will move two steps ahead and occur on a Tuesday. In 2009, it will move only a day ahead and occur on a Wednesday.

RULE C (applies to January & February)

Rule C reads as follows:

A date which falls either in the month of January or February takes a regular one step ahead as every year passes by. However, it takes two-steps ahead in a year succeeding a leap year.

Example:

4th January 2002 is a Friday. It will become a Saturday in 2003. It will become a Sunday in 2004. It will jump two steps ahead and become a Tuesday in 2005. This happens because the year 2005 succeeds the leap year 2004 and hence it moves two steps ahead.

A year succeeding a leap year can be found by adding 1 to any leap year. Thus, 2005 (2004 + 1), 2009 (2008 + 1), 2013 (2012 + 1), are years succeeding leap years.

Let us represent our findings in a tabular manner:

For January and February	For March to December
Rule A: A date moves a step ahead as every year passes by	**Rule A:** A date moves a step ahead as every year passes by
Rule C: A date moves two steps ahead in a year succeeding a leap year	**Rule B:** A date moves two steps ahead in a leap year

RULE D

A leap year can be found by dividing the last two digits of a year by 4. If the last two digits are perfectly divisible by 4, then it is a leap year.

RULE E

A century is a leap year if the first two digits are divisible by 4. For example, the years 1900, 2100, 2200, 2300 are not leap years as the digits 19, 21, 22, 23 are not divisible by 4. However, the years 2000, 2400, 2800, 3200 are leap years as the numbers 20, 24, 28 and 32 are perfectly divisible by 4.

So, these were the 5 Rules which helped us gain insights into the characteristics of dates. These 5 rules will be of great help to students and will help them to crack virtually any problem on dates. Given below are three solved examples which will be followed by the exercise.

After solving the exercise, I urge you to check the Appendix section where I have explained the 'Zeller's Rule.' This is a simple arithmetic formula that will help you to find the day on which a given date falls relating to any century. We have seen a system in this chapter by which you can find the day on which any date falls for this century only. However, 'Zeller's Rule' will enable you to calculate the day on which any date will fall for any century.

SOLVED EXAMPLES

(Q)   If 31st December 2000 is a Sunday, what day will it be on 2nd January 2005?

Ans: 31st December 2000 is a Sunday. It will fall on Monday, Tuesday and Wednesday in the years 2001, 2002 and 2003 respectively. In 2004, it will become a Friday. Hence, on 2nd January 2005 (2 days later) it will be a Sunday.

(Q)   If 17th July 2010 is a Sunday, then what day will it be on 17th July 2007 ?

Ans: In this sum we have to do backward calculation. We are given that 17th July 2010 is a Sunday. Thus, we move one step behind and find that in 2009, it was a Saturday. In

2008 it was a Friday. And in 2007 we move two steps behind and find that it was a Wednesday.

(As per rule B, a date moves two steps ahead in a leap year. Hence, we have to go two steps back from a leap year when we are going backwards.)

(Q)   If 1st January 2099 is a Thursday, what day will it be on 1st January 2101?

Ans: We are given that 1st January 2099 is a Thursday. The next year 2100 is not a leap year as the first two digits are not divisible by 4. Hence, the date will move one step ahead only and become a Friday. In the year 2101 it will move a further step ahead and become a Saturday. Thus, first January 2101 is a Saturday.

EXERCISE

PART A

Q. (1) Find which of the following years are leap years and which are not:
2000, 2100, 2101, 2040, 2004, 1004, 2404, 1404, 4404

Q. (2) Given that the key for the current year 2005 is 266315374264. Find the days corresponding to the following dates:

(i)   7th Jan

(ii) 3rd Dec

(iii) 14th Nov

(iv) 28th Aug

(v)   26th June

(vi) 30th Dec

PART B

Q. (1) Chris has provided us with the details of the birthdates
        of his family members. Find the days on which they
        were born.

        (1) Father: 1 December 1953
        (2) Mother: 4 January 1957
        (3) Grandpa: 9 December 1924
        (4) Brother: 26 January 1984

PART C

Q. (1) Given that 31st March 2002 is a Sunday, find the days
        on which the following dates will fall:

        (1) 31 March 2005
        (2) 2 April 1999
        (3) 23 March 2004
        (4) 7 April 2000

General Equations

Almost all the chapters that we have discussed till now have dealt with the arithmetical part of Vedic Mathematics. In this chapter, we will study the algebraic part of this science.

In Vedic Mathematics there are many simple formulae for solving different types of equations. Each such formula can be used for a particular category of equations. Let us assume that we have to solve an equation of the type $ax + b = cx + d$. Then the answer to this equation can be easily determined using the formula.

$$x = \frac{d - b}{a - c}$$

Let us have a look at a couple of examples:

(Q) Solve $5x + 3 = 4x + 7$.

    (a) (b) (c) (d)

The question asks us to solve the equation $4x + 7 = 5x + 3$. The equation is of the type $ax + b = cx + d$ where the values of a, b, c and d are 5, 3, 4 and 7 respectively. The value of x can be solved by using the simple formula as given below:

$$x = \frac{d - b}{a - c}$$

$$\text{Thus, } x = \frac{7 - 3}{5 - 4} = \mathbf{4}$$

(Q) Solve the equation $5x + 3 = 6x - 2$.

We represent the equation in the form of $ax + b = cx + d$ and get the values of a, b, c and d as 5, 3, 6 and -2 respectively. The value of x is

$$x = \frac{d - b}{a - c}$$

$$x = \frac{-2 - 3}{5 - 6}$$

Thus we get a simple rule that equations of the type $ax + b = cx + d$ can be solved using the formula $x = \dfrac{d - b}{a - c}$

METHOD TWO

The method given above was the simplest method of solving equations. Now we come to another method. This particular method is used to solve equations of the type $(x + a)(x + b) = (x + c)(x + d)$. The value of x will be found using the formula

$$x = \frac{cd - ab}{a + b - c - d}$$

Q. Solve the equation $(x + 1)(x + 3) = (x - 3)(x - 5)$.

The above equation is of the type $(x + a)(x + b) = (x + c)(x + d)$. Thus the numbers 1, 3, -3 and -5 are represented by the letters a, b, c and d respectively. The value of x will be determined using the formula

$$x = \frac{cd - ab}{a + b - c - d}$$

$$= \frac{15 - 3}{1 + 3 + 3 + 5}$$

$$x = 1$$

(Q) Solve the equation $(x + 7)(x + 12) = (x + 6)(x + 15)$.

We represent the numbers 7, 12, 6 and 5 with the letters a, b, c and d respectively. The value of x will be

$$x = \frac{cd - ab}{a + b - c - d}$$

$$x = \frac{90 - 84}{7 + 12 - 6 - 15}$$

$$x = -3$$

It can be inferred from the above examples, that the value of the equation $(x + a)(x + b) = (x + c)(x + d)$ can be easily determined using the principles of Vedic Mathematics.

We can tabulate our findings as follows:

Format of Equation	Value of X
$ax + b = cx + d$	$x = \dfrac{d - b}{a - c}$
$(x+a)(x+b) = (x+c)(x+d)$	$x = \dfrac{cd - ab}{a + b - c - d}$

CHAPTER **12**

Simultaneous Linear Equations

In the previous chapter we studied how to solve two types of basic equations. In this chapter we will study how to solve simultaneous linear equations. The need to solve these equations arises while cracking word-problems frequently asked in all types of exams.

What are simultaneous linear equations

Simultaneous linear equations have two variables in them. Let us say x and y. Since there are two variables in the equation we cannot solve it by itself. We need another equation with the same variable values to find the answer. When these two equations are solved together we get the values of the variables x and y.

Before studying the Vedic approach to solving simultaneous linear equations (SLE's) let us recall the traditional method of solving them.

(Q) Find the values of x and y given the equations
$2x + 4y = 10$ and $3x + 2y = 11$.

$$2x + 4y = 10 \quad\rule{3cm}{0.4pt}\quad (1)$$

$$3x + 2y = 11 \quad\rule{3cm}{0.4pt}\quad (2)$$

The co-efficients of x are 2 and 3 respectively, and the co-efficients of y are 4 and 2 respectively. In order to proceed with the solution we have to equalize either the co-efficients of x or the co-efficients of y. This can be done by multiplying the equations with suitable numbers.

In this case, we shall multiply equation (1) with 3 and equation (2) with 2. The new equations are:

$$6x + 12y = 30 \quad\rule{3cm}{0.4pt}\quad (1)$$

$$6x + 4y = 22 \quad\rule{3cm}{0.4pt}\quad (2)$$

We can see that the co-efficient of x is same in both equations. Now, we subtract the second equation from the first and get the value of y.

$$6x + 12y = 30 \quad\rule{3cm}{0.4pt}\quad (1)$$

$$\underline{6x + 4y = 22 \quad\rule{3cm}{0.4pt}\quad (2)}$$

$$8y = 8$$

We get $8y = 8$ and therefore value of y is 1. Next, we substitute the value of y in the first equation.

$$2x + 4(1) = 10$$

$$2x = 6$$

$$x = 3$$

Ans: The values of x and y are 3 and 1 respectively.

In the traditional method a new set of equations is formed in order to equalize the co-efficients of any one variable. But forming new equations is a time-consuming procedure.

Secondly, equalizing the co-efficients is not always an easy task. If the co-efficients have big numbers or decimal values it becomes very difficult to equalize them by multiplying them with suitable numbers. If the values of x are numbers like 0.5 or 0.2 or big numbers like 32 or 54 then it becomes difficult to calculate their values. In each of the two examples mentioned above it is not easy to equalize the co-efficient and involves some effort. Secondly, the possibility of making a mistake with this method is pretty high.

We will now study an alternate approach given by Vedic Mathematics.

METHOD

In this method we will not be forming new equations but instead we will calculate the values of x and y with the given equations only. The value of the variables x and y will be expressed in the form of numerator upon denominator.

$$x = \frac{\text{Numerator}}{\text{Denominator}} \qquad y = \frac{\text{Numerator}}{\text{Denominator}}$$

It should be noted that although one can find the values of both x and y there is no need for doing it. If we obtain the value of either x or y then the value of the other variable can easily be obtained by substitution. We will solve three examples by calculating the value of x and two examples by calculating the value of y.

Calculating the value of 'x'

(Q) Find the values of x and y for the equations 2x + 4y = 10 and 3x + 2y = 11.

As I said, we will calculate the value of x as numerator upon denominator. The value of numerator will be:

$$2x + 4y = 10 \underline{\hspace{5cm}} (1)$$
$$3x + 2y = 11 \underline{\hspace{5cm}} (2)$$

The numerator is obtained by cross-multiplying (4×11) and subtracting from it the cross product of (2×10) as shown by the arrows in the diagram above.

$$x = \frac{\text{Numerator}}{\text{Denominator}} = \frac{(4 \times 11) - (2 \times 10)}{\text{Denominator}} = \frac{24}{\text{Denominator}}$$

Next, we will calculate the value of the denominator.

$$2x + 4y = 10 \underline{\hspace{5cm}} (1)$$
$$3x + 2y = 11 \underline{\hspace{5cm}} (2)$$

The denominator is obtained by cross-multiplying (4×3) and subtracting from it the cross product of (2×2) as shown by the arrows in the diagram above.

$$x = \frac{\text{Numerator}}{\text{Denominator}} = \frac{24}{(4 \times 3) - (2 \times 2)} = \frac{24}{8} = 3$$

Thus, we have obtained the value of x as 3. Now, we will substitute the value of x in the equation $2x + 4y = 10$

$$2(3) + 4y = 10$$
$$6 + 4y = 10$$
$$y = 1$$

Therefore, the values of x and y are 3 and 1 respectively.

(Q) Solve the equations $2x + y = 5$ and $3x - 4y = 2$.

Numerator	Denominator
$2x + 1y = 5$	$2x + 1y = 5$
$3x - 4y = 2$	$3x - 4y = 2$
$= (1 \times 2) - (-4 \times 5)$	$= (1 \times 3) - (-4 \times 2)$
$= 22$	$= 11$
$x = \dfrac{\text{Numerator}}{\text{Denominator}} = \dfrac{22}{11} = 2$	

On substituting the value of $x = 2$ in equation (1) we get the value of y as 1. The solution set is (2, 1)

(Q) Solve the equations $32x + 12y = 120$ and $22x + 17 y = 100$.

We divide equation (1) by 4 and make it $8x + 3y = 30$ and equation (2) remains as it is.

Numerator	Denominator
$8x + 3y = 30$	$8x + 3y = 30$
$22x + 17y = 100$	$22x + 17 y = 100$
$= (3 \times 100) - (17 \times 30) = -210$	$= 66 - 136 = -70$
$x = \dfrac{\text{Numerator}}{\text{Denominator}} = \dfrac{-210}{-70} = 3$	

On substituting the value of x as 3 in equation (1) we get the value of y as 2.

We have seen three different equations wherein we calculated the value of x and substituted the value of x in a given equation to find the value of y. In the next example, we will calculate the value of y and substitute it in any one of the equations to find the value of x.

Calculating the value of y

The value of y will also be calculated in the form of numerator upon denominator. However, the technique of calculating the denominator is same as the previous technique (in case of x) and so we have to study the technique of calculating the numerator only.

(Q) Solve the equations $6x + 4y = 50$ and $5x + 5y = 50$.

Numerator	Denominator
$6x + 4y = 50$	$6x + 4y = 50$
$5x + 5y = 50$	$5x + 5y = 50$
$= (50 \times 5) - (50 \times 6)$	$= (4 \times 5) - (6 \times 5)$
$= -50$	$= -10$
$y = \dfrac{\text{Numerator}}{\text{Denominator}} = \dfrac{-50}{-10} = 5$	

In this case, we have obtained the value of y as 5. We substitute the value of y in equation (1) and get the value of x as 5. The solution set is (5, 5).

(Q) Solve $5x + 4y = 3$, $2x - 3y + 8 = 0$.

The second equation $2x - 3y + 8 = 0$ can be written as $2x - 3y = -8$. We will solve for the value of y and then substitute to find the value of x

Numerator	Denominator
$5x + 4y = 3$	$5x + 4y = 3$
$2x - 3y = -8$	$2x - 3y = -8$
$= (3 \times 2) - (-8 \times 5)$	$= 8 - (-15)$
$= 46$	$= 23$
$y = \dfrac{\text{Numerator}}{\text{Denominator}} \quad \dfrac{46}{23} = 2$	

Substituting the value of y as 2 in equation (1) we get the value of x as -1. The solution set is (-1, 2).

It can be observed that the technique for calculating the denominator is same in either method, viz. solving for x or solving for y. However, the technique of calculating the numerator is different in the second method.

When confronted with a problem, a student can calculate either the value of x or y and substitute its value in the other variable. However, one rule of thumb can be stated here which will help you in deciding which variable to solve.

If the co-efficients of x are big numbers than calculate the value of x and substitute for y and if the co-efficients of y are big numbers than calculate the value of y and substitute for x.

(This happens because when you calculate the value of x you will be dealing with the y co-efficients twice and hence avoiding the big x co-efficients and vice versa.)

SPECIFIC CASE

There is a special sutra of Vedic Mathematics called the 'Sunyam Anyat' which says 'If one is in ratio, the other is zero.' This sutra is useful when the co-efficients of either x or y are in a certain ratio.

Example:

$$5x + 8y = 40$$
$$10x + 11y = 80$$

In the above case, the x co-efficients are in the ratio of 1:2 (5:10) and the constants are also in the ratio of 1:2 (40:80). Now, our sutra says that 'if one (variable) is in ratio, the other one (the other variable) is zero.'

In this case, we see that the variable x is in ratio with the constant terms and therefore 'the other', namely, variable y, is zero. Thus, value of y is zero. The value of y can be substituted as zero in the above equation. If we take equation (1) and substitute the value of y as zero, we have $5x = 40$ and hence x $= 8$.

Example 2

$$67x + 302y = 1510$$
$$466x + 906y = 4530$$

The y co-efficients are in the ratio of 1:3 (302:906) and the constants are also in the ratio of 1:3 (1510:4530). Since the variable y is in the same ratio as the constant terms, the value of variable x is zero. We now substitute the value of x as zero in the first equation and get the value of y as 5. The values of x and y are 0 and 5 respectively.

In this example, since the co-efficient and constant terms are big numbers it would have been very difficult to calculate the answer. But, thanks to the Sunyam Anyat rule, we can easily solve them by detecting a ratio amongst the variable y.

EXERCISE

PART A

Q. (1) Solve the first three equations by calculating for x and the next three equations by calculating for y. Write the answer in the form of (value of x, value of y).

(1) $4x + 3y = 25$ and $2x + 6y = 26$

(2) $9x + 10y = 65$ and $8x + 20y = 80$

(3) $8x + 4y = 6$ and $4x + 6y = 5$

(4) $7x + 2y = 19$ and $4x + 3y = 22$

(5) $2x + 9y = 27$ and $4x + 4y = 26$

(6) $40x + 20y = 400$ and $80x + 10y = 500$

PART B

Q. (1) Solve the following equations by detecting a ratio amongst any variable:

   (1) $39x + 64y = 128$

         $63x + 128y = 256$

   (2) $507x + 922y = 1000$

         $2028x + 1634y = 4000$

PART C

Q. (1)   Solve the following word problems:

   (a) A man has one rupee and two rupee coins in his purse. The total number of coins is 52 and the total monetary value of the coins is 81 rupees. Find the number of one rupee and two rupee coins.

   (b) The monthly incomes of Tom and Harry are in the ratio of 4:3. Both of them save Rs. 800 per month. Their expenditures are in the ratio of 3:2. Find the monthly income of Tom.

   (c) The average of two numbers is 45. Twice the first number equals thrice the second number. Find the numbers.

   (d) There are two classrooms having a certain number of students. If ten students are transferred from the first classroom to the second the ratio becomes 5:9. If ten students are transferred from the second classroom to the first, the ratio becomes 1:1. Find the number of students in each classroom.

ADVANCE LEVEL

Square Roots of Imperfect Squares

In the first section, we studied how to find the square root of perfect squares. We dealt with numbers of 3 and 4 digits. In this chapter we will expand our study to include bigger numbers containing 5, 6 and more digits. We will also discuss how to find the roots of imperfect squares.

The technique mentioned in this chapter does not belong solely to Vedic Mathematics. Mathematicians have been using it as a part of their general practice. The technique for calculating square roots as described in Vedic Mathematics is different.

The method explained in this chapter is covered by some educational boards in their school syllabus. However, most students are not able to retain it for a long time. Plus there is a vast category of people who are completely unaware of it. Probably, that is the reason why they get overwhelmed when I discuss it in my seminars.

Initially I thought of dealing with the entire study of 'square roots' in a single chapter. However, the technique of general squares is slightly difficult as compared to the technique of perfect squares. Hence I decided to break the study into two different parts. The technique given in the first section of the book (Basic Section) deals only with perfect squares and the current chapter deals in general squares (perfect as well as imperfect). The technique of perfect squares as mentioned in the first section is very simple and the younger age group of readers will find it extremely useful. The technique of general squares discussed in this chapter is comparatively difficult and hence is discussed in the Advanced Section of the book.

Students giving competitive exams and professionals are urged to read this chapter thoroughly and understand its contents.

CHARACTERISTICS

- We can determine the last digit of the square root by observing the last digit of the square.

The Last Digit of the Square	The Last Digit of the Square Root
1	1 or 9
4	2 or 8
9	3 or 7
6	4 or 6
5	5
0	0

- A perfect square will never end with the digits 2, 3, 7 or 8.

- While calculating cube-roots, we divided the given number into groups of three digits each. While calculating square roots, we will divide the number into groups of two digits each starting from the right. If a single digit is left in the extreme left it will be considered a group in itself.

Example: 65536   ⟶ $\overline{6}$ $\overline{55}$ $\overline{36}$

                  996004   ⟶ $\overline{99}$ $\overline{60}$ $\overline{04}$

In the number 65536, we start from the right and take the last two digits 36 and form a group. Next, we take 55 and form a group. The single remaining digit 6 forms a group in itself.

In the number 996004, we start from the right and form a group of the digits 04. Next, we form another group of the digits 60. Finally, we form a group of the digits 99.

- The number of digits in the square root will be equal to the number of groups formed in the square. In the example above the number 65536 is converted into three groups and hence the square root will be a three-digit answer. Similarly 996004 is also converted into three groups and hence the square root will be a three digit answer.

(Note: In the case of a pure decimal number, the grouping of 2-digit numbers will take place from left to right starting from the digit after the decimal. In case of a mixed number, the grouping of numbers will take place from right to left for the integral part and from left to right for the decimal part.)

METHOD

Before solving the various examples, we will have a look at the two rules used in this system.

Rule 1: 'After every step, add the quotient to the divisor and get a new divisor.'

Rule 2: 'A new divisor can be multiplied by only that number which is suffixed to it.'

At first glance it is difficult to understand the rules. However, the examples will make them clear. We will begin with small numbers and gradually advance to big numbers.

(Q) Find the square root of 529.

$$
\begin{array}{r|rr|l}
2 & 5 & 29 & 2 \\
\hline
& -4 & & \\
\hline
& 1 & &
\end{array}
$$

- We form two groups containing the digits 5 and 29 respectively.

- We will start with the first group containing the digit 5. Try to find a perfect square just smaller to 5. A perfect square just smaller to 5 is 4 and is obtained by multiplying (2×2). Hence, we put two in the divisor column, 2 in the quotient column and write their product 4 below 5 (as we do in normal division). When 4 is subtracted from 5 the remainder is 1.

- The remainder 1 cannot be divided by 2. So, we bring down the next group of digits 29 and make the dividend 129.

$$
\begin{array}{r|rr|l}
2 & 5 & 29 & 2 \\
\hline
& -4 & & \\
\hline
4 & 1 & 29 &
\end{array}
$$

- Refer to Rule 1. It says 'add the quotient to the divisor and get the new divisor.' In this case, we add 2 (quotient) to 2 (divisor) and make it 4 (new divisor). The new divisor is 4 and the new dividend is 129.

- Refer to Rule 2. It says 'a new divisor can me multiplied by only that number which is suffixed to it.' In this case,
  If we suffix 'one' to 4 it will become 41 and $(41 \times 1 = 41)$

If we suffix 'two' to 4 it will become 42 and (42 × 2 = 84)

If we suffix 'three' to 4 it will become 43 and (43 × 3 is 129)

- If we suffix 'three' to 4 it will become 43 and the product 129 so obtained will complete the division. The remainder is zero.

```
        2  | 5   29 |23
           | -4
       ____|_____
       43  | 1   29
           | -1  29
           |_____
           |      0
```

Therefore, the square root of 529 is 23.

(Q) Find the square root of 3249.

```
        5  | 32  49 |5
           | -25
       ____|_____
           |  7
```

- We take the first group 32. The perfect square just smaller to it is 25, which is obtained by multiplying 5 with 5. We put 5 in the divisor column and 5 in the quotient column and write down 25 as the answer. The remainder is 7.

- Next, we follow Rule 1 and add the quotient 5 to the divisor 5 and make it 10.

- Note that the remainder 7 cannot be divided by 10 and hence we bring down the next group of digits 49 (from 3249) and make the new dividend as 749.

$$
\begin{array}{c|cc|l}
5 & 32 & 49 & \underline{5} \\
  & -25 & & \\
\hline
10 & 7 & 49 & \\
\end{array}
$$

- The new dividend is 749 and the new divisor is 10. Now we come to Rule 2 which says that a new divisor can be multiplied by only that number which is suffixed to it. In this case, we suffix the number 'seven' to 10 and make it 107. When 107 is multiplied by 7 the answer is 749 and the remainder is zero.

$$
\begin{array}{c|cc|l}
5 & 32 & 49 & \underline{57} \\
  & -25 & & \\
\hline
10(7) & 7 & 49 & \\
      & -7 & 49 & \\
\hline
      & & 0 & \\
\end{array}
$$

The complete procedure is exactly as given in the diagram above. The square root of 3249 is 57.

(Q) Find the square root of 65536.

- We will convert the number 65536 and write it as 6 55 36.

$$
\begin{array}{c|ccc|l}
 & 6 & 55 & 36 & \underline{} \\
\end{array}
$$

- The first group contains the digit 6. We have to find a perfect square just smaller than 6. In this case it will be 4 (2 × 2 is 4). So, we write down 2 in the divisor column and 2 in the quotient column and subtract the answer 4 from 6. The remainder is 2

$$
\begin{array}{c|ccc|l}
2 & 6 & 55 & 36 & \underline{2} \\
  & -4 & & & \\
\hline
  & 2 & & & \\
\end{array}
$$

- As per Rule 1, we will add the quotient to the divisor and get a new divisor. This addition will be repeated after every step. In this case, we have a quotient 2. We will add the quotient 2 to the divisor 2 and get a new divisor 4. Since the remainder 2 is not divisible by the new divisor 4, we will bring down the next group of digits 55.

$$
\begin{array}{r|lll|l}
2 & 6 & 55 & 36 & \underline{2} \\
  & -4 & & & \\
\hline
4 & 2 & 55 & & \\
\end{array}
$$

- As per Rule 2, we have to suffix a digit to 4 and multiply the number so obtained with the suffixed digit. Thus, we suffix the digit 'five' to 4 and make it 45. When 45 is multiplied by 5 the answer is 225. When 225 is subtracted from 255 is remainder is 30.

$$
\begin{array}{r|lll|l}
2 & 6 & 55 & 36 & \underline{25} \\
  & -4 & & & \\
\hline
4(5) & 2 & 55 & & \\
     & 2 & 25 & & \\
\hline
     & & 30 & & \\
\end{array}
$$

- Now, the new remainder is 30 and the new divisor is 45. The remainder cannot be divided by the divisor and hence we bring down the last group of digits 36 and make the dividend as 3036. We repeat rule 1 again and add the quotient to the divisor and get the new divisor. The quotient of the previous step 'five' will be added to 45 to make it 50.

$$
\begin{array}{r|lll|l}
2 & 6 & 55 & 36 & \underline{25} \\
  & -4 & & & \\
\hline
4(5) & 2 & 55 & & \\
     & 2 & 25 & & \\
\hline
50 & & 3036 & & \\
\end{array}
$$

- Once again we have to suffix a digit to the new divisor. This time we will suffix 'six' to 50 and make it 506. Since the digit 'six' is affixed, we can multiply 506 with 'six' only. The product is 3036 and the remainder is zero. Thus, the process is complete.

```
   2  |  6   55   36 | 256
      |  -4
4(5)  |  2   55
      |  2   25
50(6) |      30   36
      |     -30   36
      |            0
```

Accordingly, the square root of 65536 is 256.

- A few other examples:

```
   8  | 65   61 | 81   The square root of 6561 is 81
      | 64
161   |  1   61
      |  1   61
      |       0
```

```
   4  | 22   27   84 | 472   The square root of 222784
      | 16                    is 472
87    |  6   27
      |  6   09
942   |      18   84
      |      18   84
      |            0
```

$$\begin{array}{r|rrrr|l} 1 & 2 & 01 & 92 & 41 & 1421 \\ & 1 & & & & \\ \hline 24 & 1 & 01 & & & \\ & & 96 & & & \\ \hline 282 & & 5 & 92 & & \\ & & 5 & 64 & & \\ \hline 2841 & & 28 & 41 & & \\ & & 28 & 41 & & \\ \hline & & & & 0 & \end{array}$$

The square root of 2019241 is 1421

In all of the above-mentioned examples, we took big numbers which were perfect squares. Now, we will expand our discussion to include numbers which are imperfect squares. Their answer will be a number with a decimal value.

Q. Find the square root of 792.

The number 792 is not a perfect square. Hence, the answer will have a decimal value. However, the process of finding the square root is the same as discussed above:

$$\begin{array}{r|rr|l} 2 & 7 & 92 & 28 \\ & 4 & & \\ \hline 48 & 3 & 92 & \\ & 3 & 84 & \\ \hline 56(\ ) & & 8 & \end{array}$$

- Here we group the dividend into two parts containing the digits 7 and 92 respectively.

- We take 7 and find the perfect square just smaller to it. The perfect square is 4 which is obtained by multiplying 2 with 2. We write 2 in the divisors column and we write 2 in the quotient column and write down the answer as 4.

- The difference of 7 and 4 is 3 which is written down. The new divisor (on adding 2 to 2) is 4. Since 3 cannot be divided by 4 we bring down the next group of digits 92.

The new dividend is 392 and the new quotient is 48. We multiply 48 with 8 and write the answer as 384. The difference is 8.

- The new divisor obtained on adding 8 to 48 is 56. The new divisor is 56 and the new remainder is 8. We cannot bring down any other digits from the main dividend — 792. Hence, at this stage we simply divide 8 by 56 using normal division and get the answer as 0.14.

- The quotient already obtained is 28. The answer obtained in the previous step is 0.14. Hence, the complete answer is 28.14.

(Note: The answer obtained, 28.14, is actually an approximate answer.)

Q. Find the square root of 656.

$$
\begin{array}{r|rr|l}
2 & 6 & 56 & 25 \\
  & 4 &   &   \\ \hline
45 & 2 & 56 &   \\
   & 2 & 25 &   \\ \hline
50(\ ) & & 31 &
\end{array}
$$

In the second-to-last step, we add 5 from the quotient to the divisor 45 and get the new divisor as 50. The remainder is 31. However, with 31 as the remainder and 50 as the dividend we cannot continue the division. Hence, we divide 31 by 50 and get the answer as 0.62. But, we already have 25 in the quotient. Hence, the complete answer as 25.62 (approximate).

Q. Find the square root of 4563.

$$
\begin{array}{r|rrr|l}
6 & 45 & 63 & 67 \\
  & 36 &   &   \\ \hline
127 & 9 & 63 &   \\
    & 8 & 89 &   \\ \hline
134(\ ) & & 74 &
\end{array}
$$

The quotient already obtained is 67. When the remainder 74 is divided by 134 the answer is 0.55. Thus, the square root of 4563 is 67.55 (approximate).

From the above examples, it can be inferred that the technique for finding the square root of imperfect squares is similar to the technique of finding square roots of perfect squares. Because of a single system, our task becomes very simple as there is a need to remember just one procedure.

Now we come to the last part of this chapter where we will discuss how to find the square roots of decimal numbers.

Two important rules have to be remembered while squaring a decimal value:

Rule 1: The grouping of the integral part will be done from right to left, and the grouping of the decimal part will be done from left to right. Thus,

$$538.7041 \longrightarrow 5\ 38.70\ 41$$
$$0.055696 \longrightarrow 0.05\ 56\ 96$$
$$0.6 \longrightarrow 0.6$$

Rule 2: If there are odd number of places after the decimal, make them even by putting a zero. Thus, 0.6 will be converted to 0.60.

These are the two rules that we will keep in mind while solving the sums. The remaining steps are the same.

Q. Find the square root of 538.7041.

2	5 38.70 41	23.21
	4	
43	1 38	
	1 29	
462	970	
	924	
4641	4641	
	4641	
	0	

- The first group in the dividend is 5. The perfect square just below it is 4. We multiply 2 by 2 and write the answer and difference as 4 and 1 respectively.

- We add the quotient to the divisor and get the new divisor 4, and we bring down the next group of digits, viz. 38, and make a new dividend of 138.

- We affix 3 to 4 and make it 43. Since 3 is affixed, we can multiply 43 only with 3. The product is 129 and the difference of 138 and 129 is 9.

- We add the quotient 3 to the divisor 43 and make it 46. We bring down the next group of digits 70 and make the dividend 970.

- The quotient at this point of time is 23. However, since we have crossed the decimal part of the dividend we put a small decimal point in the quotient after 23.

- Then affix 2 to 46 and make it 462. When 462 is multiplied by 2, the answer is 924 and the difference is 46. We bring down the final group of digits 41 and affix it to 46. The new dividend is 4641. The new divisor obtained by affixing 2 to 462 is 464.

- We affix 1 to 464 and make it 4641. This number when multiplied by 1 yields itself and the process is complete with the remainder of zero.

The square root of 538.7041 is 23.21.

It can be observed from the steps that the procedure is the same as we have discussed since the beginning of the chapter. The only difference being that in this case we have to put a decimal point in the quotient when we cross the decimal part of the dividend. However, this rule is also a part of our normal division process and hence needs no elaboration.

With the study of perfect squares, imperfect squares and squares with decimal value, our study of square roots is complete. You are now geared to find the square root of any type of number.

EXERCISE

PART A

Q. (1) Find the square roots of the following perfect squares.

(1) 961

(2) 6889

(3) 12321

(4) 4084441

PART B

Q. (1) Find the square roots of the following imperfect squares.

(1) 700

(2) 1550

(3) 15641

PART C

Q. (1) Find the square roots of the following decimals (up to 2 decimal places).

(1) 0.4

(2) 150.3432

Cubing Numbers

We know that squaring is multiplying a number by itself. Cubing can be defined as multiplying a number by itself and again by itself. Thus $3 \times 3 \times 3$ is 27 and thus 27 is the cube of 3. When we cube a number we are said to have raised it to the power of 3. The cube of a number is expressed by putting a small three on the top right part of the number. For example,

$$10^3 = 1000$$
$$11^3 = 1331$$

Cubing is important while dealing with some algebraic equations and also while dealing with three-dimensional figures in geometry.

In this chapter we will study two methods of cubing numbers. Once you are well versed with both the systems, then you can decide which system to use depending on the question that you are dealing with.

METHOD ONE: FORMULA METHOD

The formula method of cubing is the first method taught to students. We have all learnt this system when we were in school. The cube of any number can be found out using the formulae:

$$(a + b)^3 = a^3 + 3a^2b + 3ab^2 + b^3$$
$$(a - b)^3 = a^3 - 3a^2b + 3ab^2 - b^3$$

Using this formulae, we were asked to express any given number as a sum or a difference of two numbers and then find its cube.

Q. Find the cube of 102.

We know that 102 is (100 + 2). We express the cube of 102 in the form of (100 + 2) where the values of a and b are 100 and 2 respectively.

$$(100 + 2)^3 = (100)^3 + 3 \times (100)^2 \times 2 + 3 \times (100) \times (2)^2 + (2)^3$$
$$= 100000 + 3 (10000) \times 2 + 300 \times 4 + 8$$
$$= 1000000 + 60000 + 1200 + 8$$
$$= 1061208$$

Q. Find the cube of 97.

The cube of 97 can be expressed as $(100 - 3)^3$.

$$(100 - 3)^3 = (100)^3 - 3 (100)^2 (3) + 3 (100) (3)^2 - (3)^3$$
$$= 1000000 - 90000 + 2700 - 27$$
$$= 912673$$

The formula method is of the simplest and most straight forward methods of cubing a number. The second method that we will study is the Anurupya Sutra of Vedic Mathematics.

METHOD TWO: THE ANURUPYA SUTRA

The Anurupya Sutra is based on the formulae that we just studied. Have a look at how it works:

$$(a + b)^3 = a^3 + 3a^2b + 3ab^2 + b^3$$

The expression on the RHS can be broken into two parts as given below. The first part has the terms a^3, a^2b, ab^2, and b^3 and the second part has the terms $2a^2b$ and $2ab^2$.

$$a^3 + a^2b + \quad ab^2 + b^3 \quad \text{————————— (1)}$$
$$+ \quad 2a^2b + 2ab^2 \quad \text{————————— (2)}$$
$$\text{Equals} \quad a^3 + 3a^2b + 3ab^2 + b^3$$

You can see that the total of the two parts equals the RHS. Now, look at the terms in the first row which are a^3, a^2b, ab^2 and b^3. Note that if we take the first term a^3 and multiply it by b/a we get the second term a^2b and if we multiply the second term a^2b by b/a we get the third term ab^2. And if the third term ab^2 is multiplied by b/a we get the final term that is b^3. Thus,

$$a^3 \times \frac{b}{a} = a^2b$$

$$a^2b \times \frac{b}{a} = ab^2$$

$$ab^2 \times \frac{b}{a} = b^3$$

On the basis of the above observations it can be concluded that as we move from left to right the terms are in geometric progression and the ratio between them is b/a. On the other hand, it can also be concluded that if we move from right to left the numbers are in geometric progression too and the ratio between them is a/b.

Now look at the second row. You will find that the numbers in the second row are obtained by multiplying the numbers above them by two. Look at the diagram given below:

$$a^3 + a^2b + ab^2 + b^3 \quad \text{———— (1)}$$

$$+ \quad 2a^2b + 2ab^2 \quad \text{———— (2)}$$

Equals $a^3 + 3a^2b + 3ab^2 + b^3$

The first term of the second row $2a^2b$ is obtained by multiplying a^2b by 2. The second term $2ab^2$ is obtained by multiplying ab^2 by 2.

In a nut-shell:

While solving the cube root of any number.

- The values of the first row can be obtained by moving in a geometric progression of b/a (from left to right) or a/b (from right to left). The first term will be a^3. When the first term is multiplied by b/a we get the second term. When we multiply the second term by b/a we get the third term. When we multiply the third term by b/a we get the fourth term. Alternatively, the fourth term can be obtained by b^3.

- The values of the second row are obtained by doubling the middle terms in the first row.

- The final cube is obtained by adding the two rows.

Whenever a number is given to us we will divide it into two parts, 'a' and 'b'. For example, if we want to find the cube of 42, then the value of 'a' will be 4 and the value of 'b' will be 2. If we want to find the cube of 74, then the value of 'a' will be 7 and 'b' will be 4. In the case of 67, then the value of 'a' will be 6 and 'b' will be 7.

We can expand the rule for three digit numbers also. If we want to find the cube of 143, the value of 'a' will be 14 and 'b' will be 3. In the case of 161, the value of 'a' will be 16 and 'b' will be 1.

Q. Find the cube of 52 (a = 5 and b = 2).

In this case the value of a is 5 and b is 2. We will move from left to right with the ratio of b/a. The value of b/a is 2/5. The terms of the first row will be as given below:

First Row

5^3 = 125
$2/5 \times 125$ = 50
$2/5 \times 50$ = 20
2^3 = 8 (alternatively we can take $2/5 \times 20$ equals 8).

Second Row

The terms of the second row is obtained by doubling the middle terms of the first row. The final figure is as given below.

	125	50	20	8
		100	40	
TOTAL:	125	150	60	8

The final answer is obtained by putting three zeros behind the first term, two zeros behind the second term, one zero behind the third term and no zero behind the last term and adding them. (After you have gone through all the examples, refer to the rule of zeros given at the end of the chapter).

Thus,

```
                    1 2 5 0 0 0
                      1 5 0 0 0
                          6 0 0
          +                   8
```

The cube of 52 is – **1 4 0 6 0 8**

Q. Find the cube of 12.

In this case, the value of 'a' is 1 and the value of b is '2'. The value of the ratio b/a is 2/1.

First Row

1^3 = 1
$2/1 \times 1$ = 2
$2/1 \times 2$ = 4
2^3 = 8 (alternatively we can take $2/1 \times 4 = 8$)

Second Row

The terms of the second row is obtained by doubling the middle terms of the first row. The final figure is as given below.

	1	2	4	8
		4	8	
TOTAL:	1	6	12	8

We put three, two, one and no zeros behind the terms respectively.

$$1\ 0\ 0\ 0$$
$$6\ 0\ 0$$
$$1\ 2\ 0$$
$$8$$
_____

Thus, the cube of 12 is – 1 7 2 8

Q. Find the cube of 31.

In this case, the value of 'a' is 3 and the value of b is '1'.

It must be remembered that we can find the cube of a number by taking the ratio b/a or a/b. When we move from left to right the ratio is b/a and when we move from right to left the ratio is a/b.

In this case we will move from right to left and the ratio will be a/b. Thus, the ratio is 3/1.

First Row

1^3 = 1
$3/1 \times 1$ = 3
$3/1 \times 3$ = 9
3^3 = 27 (alternatively we can take $3/1 \times 9$ equals 27)

Second Row

The terms of the second row are 6 and 18.

27	9	3	1
	18	6	
27	27	9	1

On adding 27000 + 2700 + 90 +1 we have 29791. Thus, the cube is 29791.

Let us have a look at a few examples:

- We will obtain the terms of the first row by geometric progression of b/a or a/b
- We will obtain the terms of the second row by doubling the middle terms
- We will add the two rows to get the final terms
- We will put three, two, one and no zeros behind the four final terms and add them to get the final answers

$$(a) \ 13^3 = \quad 1 \quad 3 \quad 9 \quad 27$$
$$\qquad\qquad\qquad\quad 6 \quad 18$$
$$\overline{\qquad\qquad\quad 1 \quad 9 \quad 27 \quad 27}$$

$$= 1000 + 900 + 270 + 27$$
$$= 2197$$

(b) $14^3 =$

1	4	16	64
	8	32	
1	12	48	64

$= 1000 + 1200 + 480 + 64$

$= 2744$

(c) $22^3 =$

8	8	8	8
	16	16	
8	24	24	8

$= 8000 + 2400 + 240 + 8$

$= 10648$

(d) $24^3 =$

8	16	32	64
	32	64	
8	48	96	64

$= 8000 + 4800 + 960 + 64$

$= 13824$

(e) $33^3 =$

27	27	27	27
	54	54	
27	81	81	27

$= 27000 + 8100 + 810 + 27$

$= 35937$

(f) $42^3 =$

64	32	16	8
	64	32	
64	96	48	8

$= 64000 + 9600 + 480 + 8$

$= 74088$

(g) $51^3 =$

125	25	5	1
	50	10	
125	75	15	1

$$= 125000 + 7500 + 150 + 1$$
$$= 132651$$

(h) $62^3 =$

216	72	24	8
	144	48	
216	216	72	8

$$= 216000 + 21600 + 720 + 8$$
$$= 238328$$

We have seen how the Anurupya Sutra works in cubing numbers of two digits. The same technique can be expanded for numbers of higher digits. We will study one example of a three digit number.

Q. Find the cube of 102.

In this case the value of 'a' is 10 and the value of 'b' is 2. We will move from left to right taking the ratio b/a (b/a = 2/10 = 1/5).

The terms of the first row are as under:

10^3 $= 1000$
$1/5 \times 1000 = 200$
$1/5 \times 200 = 40$
$1/5 \times 40   = 8$

The terms of the second row are 400 and 80.

$102^3 =$

1000	200	40	8
	400	80	
1000	600	120	8

$$= 1000000 + 60000 + 1200 + 8$$
$$= 1061208$$

We can see that the same concept can be easily applied for higher digit numbers.

The Rule Of Zeros

In the examples given above, we put 3, 2, 1 and no zeros after each step. However, this rule is applicable only up to the number 999. In other words, you can find a cube of any number up to 999 using this rule. From the number 1000 onwards we double the number of zeros that you used to put in the former case. So, from 1000 onwards we will put 6, 4, 2 and no zeros after each step. (Double of 3, 2, 1 and no zeros.) In this way the chain continues. However, we hardly come across such numbers in practical life and hence I am just providing a single example of the same without much elaboration.

Q. Find the cube of 1001.

Here, the number is above 999 and so we will put 6, 4, 2 and no zeros after the four answers obtained.

We break the number 10/01 in two parts. The value of 'a' is 10 and the value of 'b' is 1. We move from left to right and the ratio b/a is 1/10.

1000	100	10	1
	200	20	
1000	300	30	1

$$= 1000000000 + 3000000 + 3000 + 1$$
$$= 1003003001$$

In this chapter we have seen the formula method and the Anurupya Method of cubing numbers. When you are giving your exam, observe the digits used in the question and select the system which will give you the answer instantly. If the number is made up of ones, two, threes or similar digits, then use the Anurupya Sutra. If the number is closer to the bases then use the formula method.

EXERCISE

PART A

Q. (1) Find the cube of the following numbers using the formula for $(a + b)^3$.

(1) 105

(2) 41

(3) 54

(4) 23

Q. (2) Find the cube of the following numbers using the formula for $(a - b)^3$.

(5) 49

(6) 90

(7) 199

(8) 96

PART B

Q. (1) Find the cube of the following numbers using the Anurupya Rule.

(1) 66

(2) 77

(3) 91

(4) 19

PART C

Q. (1) Find the cube of the following numbers using the Anurupya Rule.

(1) 43

(2) 72

(3) 101

CHAPTER 15

Base Method of Division

The study of division is divided in two parts. In this chapter we will study the Base Method and in the next chapter we will study the Paravartya Method of division.

We have seen the applications of the Base Method in multiplication and squaring of numbers. In this method we use power of ten as a base and then calculate the difference between the base and the given number. The same concept will be used for division.

FORMAT

The style of presentation used in this system is very much unlike the traditional system. Have a look at the diagram below:

Base

Dividend

Divisor
Difference

Quotient   Remainder

We will divide the dividend in two parts. The RHS will contain as many digits as the number of zeros in the base. The final answer obtained on the LHS is the quotient and RHS is the remainder.

PROBLEMS

Q. Divide 23 by 9.

We are asked to divide 23 by 9. The divisor is 9, the base is 10 and the difference is 1.

Next, since the base 10 has one zero in it, we divide the dividend in such a way that the RHS has one digit.

We now bring down the first digit of the dividend, viz. 2, as shown in the diagram below:

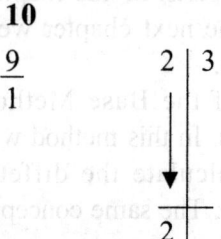

Next, we multiply 2 with the difference 1 and add the answer to the next digit of the dividend as shown below:

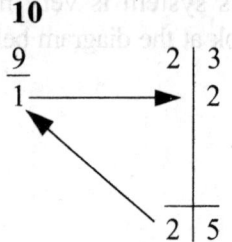

The product of 2 and 1 (the difference) is 2 which is written below 3. The sum of 3 and 2 is 5.

Thus, when 23 is divided by 9 the quotient is 2 and the remainder is 5.

Q. Divide 31 by 9.

The divisor is 9, the base is 10 and the difference is 1. We divide the dividend in two parts with one digit in the RHS as there is one zero in the base.

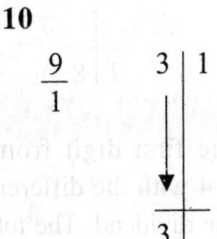

We write down the first digit of the dividend 3 as shown above.

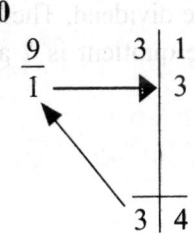

Next, we multiply the 3 written down with the difference 1. The product so obtained is written below the second digit of the dividend.

Finally, we add the digits 1 and 3 to get the answer 4.

Thus, the quotient is 3 and the remainder is 4.

Examples:

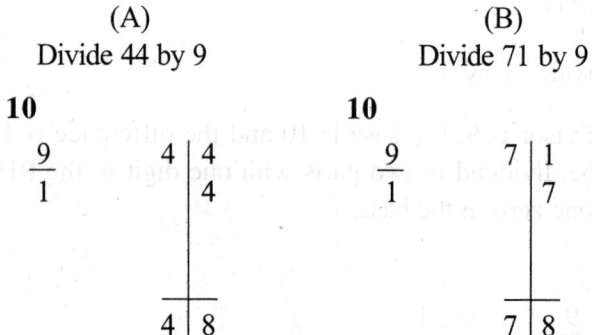

(A)
Divide 44 by 9

(B)
Divide 71 by 9

In example (A) we bring down the first digit from the dividend, viz. 4. Next, we multiply the 4 with the difference 1 and write it below the second digit of the dividend. The total of the RHS is 8. Thus the quotient is 4 and the remainder is 8.

In example (B) we bring down the first digit of the dividend, viz. 7. Next, we multiply the 7 with the difference 1 and write the answer below the second digit of the dividend. The total of RHS is 1 plus 7 equal to 8. Thus, the quotient is 7 and the difference is 8.

Q. Divide 31 by 8.

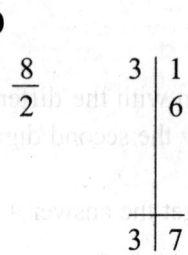

- In this case the dividend is 31 and the divisor is 8. We divide 31 into two parts with the right hand side having as many digits as the number of zeros in the base. Since, the base ten has one zero we have one digit in the RHS of the dividend.

- We write the divisor as 8 and the difference below it as 2.
- We bring down the first digit of the dividend 3 as it is.
- We multiply the 3 with the difference 2 and get the product 6. This is written down below the second digit of the dividend, viz. 1. The total is 7. Hence, the quotient is 3 and the remainder is 7.

More examples:

(A)

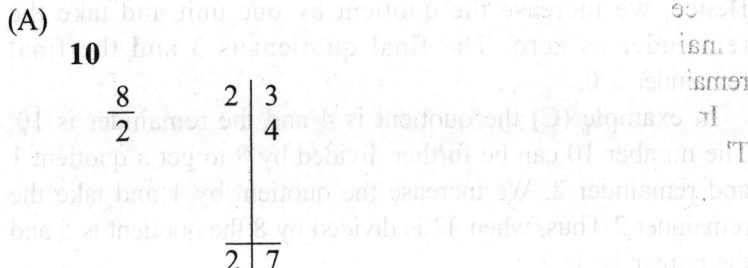

10

$$
\frac{8}{2} \quad \begin{array}{c|c} 2 & 3 \\ & 4 \end{array}
$$

$$
\begin{array}{c|c} 2 & 7 \end{array}
$$

(B)

Divide 24 by 8

10

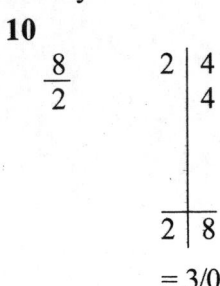

$$
\frac{8}{2} \quad \begin{array}{c|c} 2 & 4 \\ & 4 \end{array}
$$

$$
\begin{array}{c|c} 2 & 8 \end{array}
$$

$$= 3/0$$

(C)

Divide 42 by 8

10

$$
\frac{8}{2} \quad \begin{array}{c|c} 4 & 2 \\ & 8 \end{array}
$$

$$
\begin{array}{c|c} 4 & 10 \end{array}
$$

$$= 5/2$$

In example (A) we bring down the first digit 2. Next, we multiply 2 with the difference, viz. 2. The total is 4 and is written below the second digit of the dividend. The quotient is 2 and the remainder is 7.

In example (B) we bring down the first digit 2. Next, we multiply 2 with 2 and get the answer 4. The total of RHS is 8. In this case, we have the quotient as 2 and the remainder is 8. But the remainder 8 so obtained is itself equal to the divisor. Hence, we increase the quotient by one unit and take the remainder as zero. The final quotient is 3 and the final remainder is 0.

In example (C) the quotient is 4 and the remainder is 10. The number 10 can be further divided by 8 to get a quotient 1 and remainder 2. We increase the quotient by 1 and take the remainder 2 Thus, when 42 is divided by 8 the quotient is 5 and the remainder is 2.

BIGGER DIVISORS

Q. Divide 502 by 99.

100

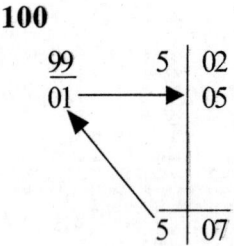

The quotient is 5 and the remainder is 7.

Q. Divide 617 by 95.

100

```
    95      6 | 17
    05        | 30
              |
              |
           6  | 47
```

In this case we write down the first digit of the dividend as it is. We multiply 6 with the difference 5 and write the answer 30 below the RHS. The total of RHS is 47. Thus, the quotient is 6 and the remainder is 47.

In all the examples that we have seen above, we split the dividend into LHS and RHS. However, there was only one digit in the LHS. We will now take a look at how to solve examples where the dividend is big and the LHS has more than one digit.

Q. Divide 123 by 9.

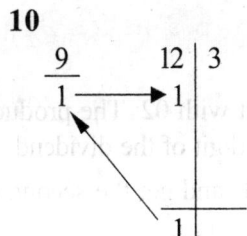

- The base 10 has one zero and therefore we split the dividend in such a way that the RHS has one digit.
- Now we are left with two digits on the LHS.
- We bring down the first digit 1 as it is.
- We multiply the 1 with the difference 1 and put the answer below the second digit of the dividend.
- The second digit of the dividend is 2 and we add 1 to it. The total is 3.

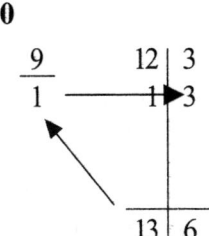

- We now multiply the total 3 with the difference 1 and write the product, viz. 3 below the third digit of the dividend.
- The total is 6.
- Thus, the quotient is 13 and the remainder is 6.

Q. Divide 1234 by 98.

100		12	34
98		0	2
02			04
		---	---
		12	58

- We bring down 1 and multiply it with 02. The product 02 is written down from the second digit of the dividend.
- We now add 2 plus 0 downwards and get the second digit of the quotient. The answer is 2.
- We multiply 2 with 02 and the final answer 04 is written down from the third digit of the dividend.
- We add up the numbers on the RHS from the extreme right column. The total of 4 + 4 is 8. We come to the column on the left and add 3 + 2 + 0 = 5.
- Hence, the product is 12 and the remainder is 58.

(In the above example, the difference of 100 and 98 is taken as 02. If we take the difference as 2 instead of 02 we will get incorrect answer.)

Q. Divide 2122 by 97.

100			
97		21	22
03		0	6
			03
		---	---
		21	85

- We bring down the first digit of the dividend — 2.
- 2 multiplied by the difference 03 gives 06 which is written just below the second digit of the dividend.
- We now add 1 plus 0 downwards and get the second digit of the quotient as 1.
- We now multiply 1 with 03 and get the answer 03, which is written below the third digit of the dividend.
- The quotient is 21 and the remainder (obtained by adding up the values in RHS) is 85.

Q. Divide 12311111 by 99970.

100000

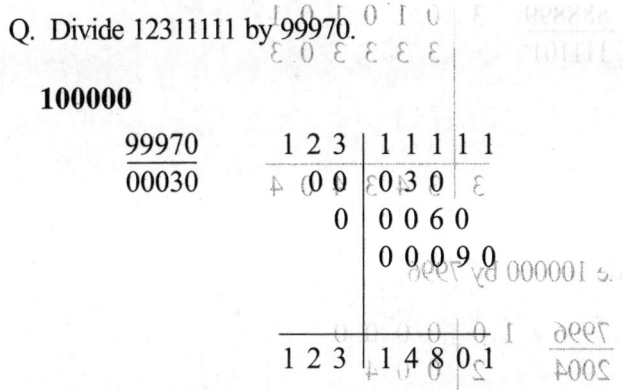

In this case we have a three digit quotient. The numbers 1, 2 and 3 are successively multiplied with 00030 to get the final answer. The quotient is 123 and the remainder is 14801.

MISCELLANEOUS EXAMPLES

(a) Divide 1212 by 88

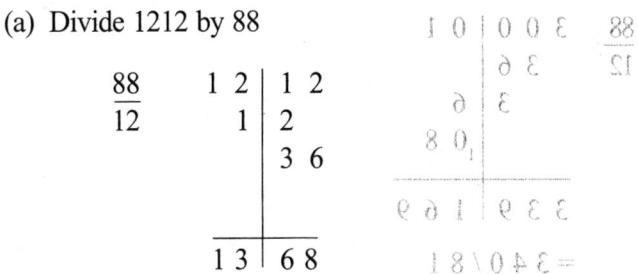

(b) Divide 112 by 79

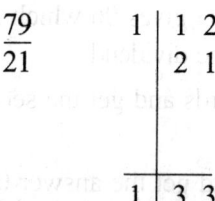

$$\frac{79}{21} \quad \begin{array}{c|cc} & 1 & 1 \ 2 \\ & & 2 \ 1 \\ \hline & & \\ & 1 & 3 \ 3 \end{array}$$

(c) Divide 3010101 by 888899

$$\frac{888899}{111101} \quad \begin{array}{c|cccccc} 3 & 0 & 1 & 0 & 1 & 0 & 1 \\ & 3 & 3 & 3 & 3 & 0 & 3 \\ \hline & & & & & & \\ 3 & 3 & 4 & 3 & 4 & 0 & 4 \end{array}$$

(d) Divide 100000 by 7996

$$\frac{7996}{2004} \quad \begin{array}{cc|cccc} 1 & 0 & 0 & 0 & 0 & 0 \\ 2 & & 0 & 0 & 4 & \\ & & 4 & 0 & 0 & 8 \\ \hline & & & & & \\ 1 & 2 & 4 & 0 & 4 & 8 \end{array}$$

(e) Divide 30001 divided by 88

$$\frac{88}{12} \quad \begin{array}{ccc|cc} 3 & 0 & 0 & 0 & 1 \\ & 3 & 6 & & \\ & & 3 & 6 & \\ & & & {}_1 0 & 8 \\ \hline 3 & 3 & 9 & 1 & 6 & 9 \end{array}$$

$$= 3 \ 4 \ 0 \ / \ 8 \ 1$$

(f) Divide 210021 by 8888

```
8888    2 1 | 0 0 2 1
1112      2 | 2 2 4
              3 3 3 6

          _____
        2 3 | 5 5 9 7
```

(g) Divide 20407 by 8987

```
8987    2 | 0 4 0 7
1013      | 2 0 2 6

          _____
        2 | 2 4 3 3
```

(h) Divide 11007 by 799

```
799    1 1 | 0 0 7
201      2 | 0 1
             6 0 3

         _____
       1 3 | 6 2 0
```

= 13 / 620

(In examples (g) and (h) there is a carry-over involved on the RHS.)

(i) Divide 2211 by 88

```
88    2 2 | 1 1
12      2 | 4
            4 8

        _____
      2 4 | 9 9
```

= 25 / 11

(j) Divide 111301 by 897

$$
\begin{array}{r|rrr|rrr}
897 & 1 & 1 & 1 & 3 & 0 & 1 \\
\hline
103 & & 1 & 0 & 3 & & \\
& & & 2 & 0 & 6 & \\
& & & & 3 & 0 & 9 \\
\hline
& 1 & 2 & 3 & 9 & 7 & 0 \\
\end{array}
$$

= 124 / 73

(k) Divide 30122 by 87

$$
\begin{array}{r|rrr|rr}
87 & 3 & 0 & 1 & 2 & 2 \\
\hline
13 & & 3 & 9 & & \\
& & & 3 & 9 & \\
& & & & {}_1 6 & 9 \\
\hline
& 3 & 3 & {}_1 3 & 2 & 8 & 1 \\
\end{array}
$$

= 343 / 281

= 346 / 20

In example (i) the divisor is 88 and the remainder is 99. However, we cannot have the remainder greater than the divisor. Therefore, we increase the value of the quotient by 1 and take the difference as the new remainder (99 - 88).

In examples (g) and (h) there is a carry-over involved on

In example (j) the divisor is 897 and the remainder is 970. The remainder is greater than the divisor. We increase the value of the quotient by 1 and take the difference as the new remainder (970 - 897).

In example (k) the total of the third quotient digit is 13. We therefore carry over the 1 to the second quotient digit. The final answer is 343. Next, we will divide the remainder by the divisor and get the new quotient as 346 and new remainder as 20.

EXERCISE

PART A

(a) Divide 102 by 74

(b) Divide 10113 by 898

(c) Divide 102030 by 7999

(d) Divide 1005 by 99

PART B

(a) Divide 431 by 98

(b) Divide 10301 by 97

(c) Divide 12000 by 889

(d) Divide 111099 by 8987

(e) Divide 30111 by 87

Division (Part Two)

In the previous chapter we saw how the Base Method can be used in the process of division. One of the drawbacks of the Base Method of division is that we can only use higher numbers like 7, 8 and 9 in the divisor. The obvious question that arises is how to solve a problem of division where the divisor includes numbers like 1, 2, 3, etc. The answer is given by the Paravartya Sutra of Vedic Mathematics, which we shall study in this chapter. The Vedic sutra 'Paravartya Yojayet' means transpose and apply.

The format and working of this system is the same as explained in the previous chapter. However, in this case we will have a negative difference.

(a) Divide 3966 by 113

$$
\begin{array}{c|cc|cc}
\mathbf{100} & & & & \\
\hline
113 & 1 & 2 & 9 & 6 \\
\text{-(13)} & & -1 & -3 & \\
& & & -1 & -3 \\
& & & & \\
\hline
& 1 & 1 & 5 & 3 \\
\end{array}
$$

- The divisor is related to the base 100 and therefore we split the dividend in such a way that the RHS has two digits.

- The base is 100 and the difference is -13 (negative).

- We write down the first digit 1 of the dividend as it is.

- We multiply 1 with the difference -13 and write the answer as -1 and -3 below the second and third digits of the dividend.

- Next, we go to the second column of the dividend. We bring down 2 minus 1 is 1.

- We multiply 1 with -13 and write the answer as -1 and -3 below the last two digits of the dividend.

- Thus, the quotient is 11 and the difference is 53.

(Note: In this case we have represented the difference as -13. Alternatively, it can be shown as -1-3. From the second example onwards we will use the latter way.)

(b) Divide 2688 by 120

100

$$
\begin{array}{c|cc}
120 & 2\ 6 & 8\ 8 \\
\text{-2-0} & \text{-4} & \text{-0} \\
 & & \text{-4-0} \\
\hline
 & 2\ 2 & 4\ 8 \\
\end{array}
$$

- The divisor is 120. Therefore, the base is 100 and the difference is -2-0.

- We bring down 2 from the dividend.

- We multiply 2 with -2-0 and write the answer as -4-0.

- Next, we go to the second digit of the dividend.

- We bring down 6 - 4 = 2.

- We multiply 2 with -2-0 and get the answer as -4-0.

- The quotient is 22 and the remainder is 48.

Examples

(c) Divide 113968 by 1023

```
   1023  1 1 3 | 9  6  8
  -0-2-3  -0 -2  -3
           -0  -2 -3
              -0 -2 -3
  _____
         1 1 1 | 4  1  5
```

Quotient = 111; Remainder = 415

(d) Divide 1999 by 180

```
   180    1 9 | 9  9
  -8-0    -8   -0
              -8 -0
  _____
          1 1 | 1  9
```

Quotient = 11; Remainder = 19

Thus we can see that the process used in this system of division is very much similar to the process that we observed in the previous chapter. Let us now have a look at a variety of different examples and the technique employed in each case.

At times we might have a negative answer in the quotient.

(e) Divide 14189 by 102

```
   102   1 4 1 | 8  9
  -0-2    -0 -2
              -0 | -8
                 | 0  2
  _____
       1 4(-1) | 1  1
```

= 139 / 11

- In this case we bring down the first digit 1. We multiply 1 with -0 -2 and write the answer below the second digit as -0-2.

- Next we bring down 4-0 =4. We multiply 4 by -0-2 and write the answer as -0-8.

- Next we bring down 1 -2 - 0 = -1 and multiply it with -0 -2 and get the answer as 0 2. (Since both the quotient and divisor are negative the answer will be positive.)

- The quotient is 140 minus 1 equals 139. The remainder is 11.

(f) Divide 110999 by 1321

$$
\begin{array}{r|rrr}
1321 & 1\ 1\ 0 & 9 & 9 & 9 \\
\text{-3-2-1} & \text{-3 -2} & \text{-1} & & \\
& 6 & 4 & 2 & \\
& & \text{-12 -8 -4} & & \\
\hline
1\ \text{-2}\ 4 & 0 & 3 & 5 \\
\end{array}
$$

= 84 / 035

Here the quotient is 100 - 20 + 4 equals 84. The remainder is 035.

At times we might have a negative answer in the remainder.

(g) Divide 1693 by 131

$$
\begin{array}{r|rr}
131 & 1\ 6 & 9 & 3 \\
\text{-3-1} & \text{-3} & \text{-1} & \\
& & \text{-9 -3} & \\
\hline
1\ 3 & \text{-10} & \\
\end{array}
$$

= 12 / 121

In example (g) the quotient is 13 and the remainder is -10. Therefore, we reduce the quotient by 1 and subtract the remainder from the divisor. Hence, the quotient is 13 -1 = 12 and the remainder is 131 - 10 = 121.

Let us understand the logic of this calculation.

Suppose we have to divide 890 by 100. Now, the quotient we have is 9 and the remainder is -10 (because 100 multiplied by 9 minus 10 is 890).

$$Q = 9 ; \quad R = -10$$

Another way of representing the number 890 is quotient = 8 and remainder = 90.

$$Q = 8 \quad R = 90$$

In this case we have reduced the quotient by 1 and reduced the remainder from the divisor. This same concept has been applied in example (g).

(h) Divide 14520 by 111

```
  111     1 4 5 | 2 0
 -1-1     -1 -1 |
              -3 | -3
                 | -1 -1
          ───────┼────────
          1 3 1  | -2 -1

        = 130 / 90
```

Here the quotient is 131 and the remainder is = -21. We cannot have a negative remainder in the final answer. Hence, we reduce the quotient by 1 and subtract the remainder from the divisor. The final quotient is 130 and final remainder is 90.

(i) Divide 16379 by 1222

$$
\begin{array}{r|rrr}
\underline{1222} & 1\ 6 & 3 & 7 & 9 \\
-2\text{-}2\text{-}2 & -2 & -2 & -2 \\
 & & -8 & -8 & -8 \\
\hline
& 1\ 4 & -7 & -3 & 1
\end{array}
$$

= 13 / 493

Here the quotient is 14 and the remainder is (-700 -30 + 1), which equals -729. We now reduce the quotient by 1 and subtract the remainder from the divisor. The final quotient is 13 and remainder is (1222 - 729) equals 493.

(b) Substitution Method

Sometimes the divisors are such that it is difficult to calculate the answer by itself. In these cases, we substitute the divisor using another number and then calculate the answer.

(j) Divide 10030 by 827

In this case the dividend is 10030 and the divisor is 827. We will solve the question using the normal method and the substitution method. In the normal method we will take the divisor as 827 and the difference as 173. In the substitution method, we will take the divisor as 827 and the difference 173 will be represented as $200 - 30 + 3 = 2 - 3 + 3$

Normal Method		**Substitution Method**	

$$
\begin{array}{r|rrr}
\underline{827} & 1\ 0 & 0 & 3 & 0 \\
173 & 1 & 7 & 3 \\
 & & 1 & 7 & 3 \\
\hline
& 1\ 1 & 9 & 3 & 3
\end{array}
\qquad
\begin{array}{r|rrr}
\underline{827} & 1\ 0 & 0 & 3 & 0 \\
2\text{-}33 & 2 & -3 & 3 \\
 & & 4 & -6 & 6 \\
\hline
& 1\ 2 & 1 & 0 & 6
\end{array}
$$

= 12 / 106

The answer is the same in either case.

(k) Divide 10000 by 819

In the normal method we will take the divisor as 819 and the difference as 181. In the substitution method we will write the divisor as 819 and the difference 181 as 200 - 20 + 1 = 2 -2 +1

<table>
<tr><td colspan="2">Normal Method</td><td colspan="2">Substitution Method</td></tr>
</table>

```
    819    1 0 | 0 0 0          819    1 0 | 0 0 0
    181      1 | 8 1           2-21    -2 |-2 1
                | 1 8 1                     | 4 -4 2
               _____               _____
            1 1| 9 9 1              1 2| 2 -3 2
```

= 12 / 172   = 12 / 172

(Remainder 200 - 30 + 2 = 172)

In the two examples given above we substituted the difference with some other number. Another way of substitution is by dividing/multiplying the divisor with a suitable number so that it becomes closer to a base and then we can multiply it quickly.

(l) Divide 1459 by 242

We divide 242 by 2 and make it 121. We will now perform the division with the new dividend 121.

```
    121    1 4 | 5 9
   -2-1     -2 |-1
                | -4 -2
              _____
       2) 1 2 | 0 7
```

= 6 / 07

The quotient is 12 and the remainder is 07. But this answer is with respect to the divisor 121. We want to find the answer with respect to the divisor 242. Since 242 is divided by 2 to obtain 121, we divide the quotient by 2 and get the answer 6. The remainder always remains the same.

(m) Divide 1112 by 33

In this case we multiply the divisor 33 by 3 and make it 99. Note that the difference in this case is 01 and not 1.

```
  99      1 1 | 1 2
 ____        |
  01      0 | 1
             | 0 1
          ___|_____
          1 1 | 2 3
       ×     3
          _____
          3 3
```

The quotient is 11 and the difference is 23. Since, 33 is multiplied by 3 to obtain 99, we multiply 11 by 3 and make it 33. The remainder remains the same. The final quotient is 33 and the final remainder is 23.

(n) Divide 12657 by 791

We divide 791 by 7 and get the answer as 113. Therefore, the new divisor is 113.

```
  113    1 2 6 | 5   7
 -1-3       -1 -3 |
            -1 | -3
               | -2 -6
          _____|_____
  7) 1 1 2 | 0   1
     = 16 / 01
```

The divisor 791 is divided by 7 to get the answer 113. Therefore, we divide 112 by 7 and get the answer as 16.

(o) Divide 1389 by 61

We multiply 61 by 2 and get the new divisor as 122.

```
    122       1 3 | 8 9
   -2-2        -2 | -2
                  | -2 -2
                  |
             1 1  | 4 7
          ×    2
             2 2 / 4 7
```

The quotient is 22 and the remainder is 47.

EXERCISE

PART A

(a) Divide 1389 by 113

(b) Divide 145516 by 1321

(c) Divide 136789 by 12131

(d) Divide 246406 by 112

PART B

(a) Divide 13592 by 114

(b) Divide 25430 by 1230

(c) Divide 15549 by 142

(d) Divide 101156 by 808 (Hint: Take difference 192 as $2 - 1 + 2$)

PART C

(a) Divide 4949 by 601 (Hint: Use $601 \times 2 = 1202$ as divisor)

(b) Divide 14799 by 492 (Hint: Use $492/4 = 123$ as divisor)

Tips for Competitive Exams

In this chapter, I will give you some instant tips that you can use for competitive exams. In exams like CAT, SAT, GMAT, GRE, UPSC, Railways, Defence, Bank PO and many other exams, a lot of emphasis is laid on quick calculation and accurate estimation of numerical problems. In many cases, we have observed that the average time per question is less than 60 seconds and so it becomes extremely important to develop the skill to quickly tackle questions...

Let us have a look at a few such handy techniques:

(A) MENTAL CALCULATION OF NUMBERS

Ever since school we are taught to do all calculations from right to left direction. Suppose you want to add up 4639 + 1235, here's how you would do the calculation.

$$4\ 6\ 3\ 9$$

$$+\ \underline{1\ 2\ 3\ 5}$$

First you would begin from the extreme right and add up 9 + 5 is 14. Then you would write 4 and carry over 1. Next you would add (3 + 3) is 6 plus 1 carried over is 7 and so on...

This technique of going from right to left works fine on paper. However, whenever you are forced to do mental calculations, a much better strategy is to move from 'left-to-right' instead of right-to-left.

Here's how you should do the above calculation.

Keep the first number 4639 as it is in your mind. Break up the second number 1235 as 1000 + 200 + 30 + 5. Now add all the numbers one after another

4639 + **1000** = 5639

This 5639 + **200** = 5839

5839 + **30** = 5869

And 5869 + **5** = 5874

Thus your final answer is 5874.

At first glance, it appears cumbersome. But once you get habituated with left-to-right calculations, you will be able to calculate with amazing speed!

The same technique can be applied for other mathematical calculations also. Let us see subtraction.

(Q) Subtract 4142 from 7580.

Now you have two possible options to solve this question. One is the traditional right-to-left method where you try to subtract 2 from 0 and realize that it is not possible. So you ask 0 to borrow from its neighbor and so on...

However, this method is unnecessarily cumbersome. Instead, if we represent the same number in left-to-right manner, we notice it becomes utmost easy to solve it.

7580 minus 4142 (**4000 + 100 + 40 +2**)

So first we have 7580 – **4000** = 3580

Next, 3580 – **100** = 3480

3480 – **40** = 3440

And finally, 3440 – **2** = 3438

When I was a young school boy, I always hated learning multiplication tables. However, I was equally fascinated by an old uncle in my neighbourhood who always boasted that he knew all multiplication tables from 1 to 100. If you would ask him what is 76 times 7 ? In a jiffy, he would answer that it is 532. Or if you would ask him what is 83 times 8, within a fraction of a second he would tell you the answer 664. During my early school years, I was deeply impressed by the fact that he had memorized so many multiplication tables. But, a few years later, he revealed to me that he never knew multiplication tables upto 100; he only knew tables upto 10. What he was doing was breaking the number from left-to-right (just as I just mentioned above) Here's how he used to do it...

Suppose you would ask him what is 76 times 7,

He would mentally break up the number 76 as 70 + 6 and then multiply each of these values by 7,

$$70 + 6$$
$$\downarrow \quad \downarrow$$
$$\underline{7 + 7}$$

- 70 x 7 is 490
- 6 x 7 is 42
- 490 + 42 is same as adding 490 + (10 + 32)
- So you first add 490 + 10 and get a convenient number of 500 and then simply add 32 to get 532

Suppose you have to multiply 83 with 8 mentally...

- First you multiply 80 with 8 and get the answer 640
- Then you multiply 3 with 8 to get 24
- The final answer (640 + 24) can be instantly written as 664.

So, please remember, whenever you have to make mental calculations, avoid the traditional method of going from right to left and instead go from left to right to get the answer quickly.

(B) Estimation of Imperfect Square Roots

Now we move on to the second technique in this series called 'Estimation of Square Roots'. Kindly please note that I have used the word 'estimation' of square roots and not used the word 'calculation of square roots'. The reason is very simple. This technique will only give you a rough idea (estimation) of the answer and not the accurate answer.

Let us see how this technique works:

(Q) Find the square root of 70.

- First we have to find a perfect square root less than 70. Hmm....so this is how we do it. We start counting downwards from 70 and come to 64 and we know that 64 is a perfect square whose square root is 8.
- Now we divide 70 by this 8 and we will get the answer (70 divided by 8) is 8.75.
- We now take the average of the two underlined numbers 8 and 8.75 to get the answer 8.37. Thus 8.37 is the approx square root of 70 (take only 2 decimal places).

(Q) Find the square root of 150.

- The perfect square just below 150 is 144 whose square root is 12.
- We divide 150 by 12 to get the answer 12.5.
- Finally, we take the averages of 12 and 12.5 to get 12.25 which is the approx. square root of 150.

(Q) Find the square root of 8200.

- The perfect square just below 8200 is 8100 whose square root is 90.
- 8200 divided by 90 gives 91.11.
- The average of 90 and 91.11 = 90.55 which is our approx. answer.

(C) Fractions, Percentages and Decimals

I have traveled across many countries of the world to conduct my workshops. During these tours I have noticed a vast difference in the mathematical aptitude of people. In some countries I have seen modestly educated shop-keepers easily calculating percentages and decimals on their finger tips while in some places I have seen people perspiring (literally!) to mentally calculate 186 plus 146.

In this sub-section, I will urge the reader to memorize this key of very important standard fractions which will help him in calculating such questions with ease.

Number	Fractional Value	Decimal Value
With 2	½	0.5
With 3	1/3	0.33
	2/3	0.67
With 4	1/4	0.25
	2/4	0.5
	3/4	0.75
With 5	1/5	0.2
	2/5	0.4
	3/5	0.6
	4/5	0.8
With 6	1/6	0.16
	2/6	0.33
	3/6	0.5
	4/6	0.67
	5/6	0.83

Number	Fractional Value	Decimal Value
With 7	1/7	0.142857
(Note how the	2/7	0.285714
digits 142857	3/7	0.428571
repeat in cyclic	4/7	0.571428
order)	5/7	0.714285
	6/7	0.857142
With 8	1/8	0.125
	2/8	0.250
	3/8	0.375
	4/8	0.500
	5/8	0.625
	6/8	0.750
	7/8	0.875
With 9	1/9	0.11
	2/9	0.22
	3/9	0.33
	4/9	0.44
	5/9	0.55
	6/9	0.66
	7/9	0.77
	8/9	0.88

The fractional and decimal value with numbers 2,3,4,5 are very easy and almost everybody knows them. The ones with 6 will require a little learning. The values with 7 repeat themselves in cyclical order. And, as can be seen by mere observation, the values with 8 increase by 0.125 as we move on and the values with 9 move by 0.11 as we move on.

It is virtually inevitable for those giving competitive exams to remember this key. (With number 7, you don't have to remember upto 6 decimal places, even 2 decimal places will be sufficient).

So if someone asks you what is 40% of 250, then, immediately on hearing the term 40% the term 2/5 should flash in your mind and you must mentally do the calculation (250 x 2 is 500 and 500 divided by 5) to get the answer 100.

If someone asks you what is the net price after 3/5th discount on 1600, then it should immediately flash in your mind that 3/5th is 60% and 60% of 1600 is 960 (because 16 times 6 is 96). And then, instead of subtracting 960, you can first subtract 1000 and then add back 40.

So, 1600 minus 1000, is 600 and when you add back 40 to 600 you get 640.

(Or better still, balance amount after deducting 3/5th discount from 1600 should be same as 2/5th of 1600. And 2/5th of 1600 is 40% of 1600 which is 640 (because $16 \times 4 = 640$).

Have you seen the way Sachin Tendulkar plays a super fast delivery of Shoaib Akhtar or Brett Lee ? The speed-o-meter says that an express delivery from someone like Shoaib Akhtar takes less than one second to reach Sachin Tendulkar's bat. Now imagine the situation, within one second Tendulkar's eyes have to see the ball, his hands and legs have to adjust themselves and his bat must come down at the right time to hit the delivery. And all this must happen within a second! Technically, it sounds difficult but ace batsmen like Sachin Tendulkar, Brian Lara and others have developed an almost 'reflex action' sort of response and so their bodies and minds work very fast to respond do the express deliveries.

The secret of cracking math and numerical questions in competitive exams is to develop this similar reflex-action sort of behavior. The moment you see a question related to numbers, within a second the numbers and calculations must start happening in your mind. For example, suppose you are solving a question where you have to find the average of 75,72 and 70. The traditional method requires you to first add 75+72+70 and then divide the total by 3 to get the average. However, if you want to crack competitive exams, you need to have a more instinctive, reflex-action sort of approach. This is how your mind should work...

'Hmm, so I have to find the average of 75, 72 and 70

Now if 75 gives 2 to 70, the numbers will be 73, 72 and 72

So the average is 72 and 1 extra.

If I divide this 1 extra by number 3, it should yield 0.33

Thus the final average is 72.33'

Although I have taken more than 5 lines to explain this question, in your mind, the entire procedure must take place in 2 to 3 seconds.

It must be remembered, that the success rate in any competitive exams is sometimes as low as 2 to 3%. In other words, out of every 100 students appearing for an exam, only 2 or 3 will make it to the next level. So in order to beat the competition, not only should your performance be super good it must also be super fast!

This is precisely the reason why I am insisting that instead of sitting and calculating each number in the traditional, sequential, line-by-line manner you must slowly cultivate the art of jumbling around with the number and get the answer in a jiffy!

Happy Solving!

Afterword

Since the time people started formally teaching Vedic Mathematics, it has spread like wildfire. Millions of people have been literally tantalized with the workings of these systems. Institutions which have invited me to conduct seminars on this subject have invited me again and again to conduct more sessions and to address larger audiences. Many institutions providing training in competitive exams to students have made teaching of Vedic Mathematics compulsory. There is literally a craze in the student community to learn these systems.

Whereas these systems are being taught informally to the students of schools and colleges, there has been rising pressure to introduce these systems formally as a part of the school and college curriculum. Many people ask me why schools and colleges don't officially conduct training programs on this subject. If these subjects are being taught formally, then it will be easy to address larger audiences and ensure that the subjects are thoroughly learnt.

Many universities in the United States, United Kingdom, Germany, Canada and other western countries have increasingly encouraged further research on these systems. They have employed faculties and trainers who regularly teach students. Everyday I hear news of some educational institution in the west which has formally accepted the study of Vedic Mathematics in its curriculum.

However, the sad thing is that our own people show resentment against the system. Whatever criticism this subject has faced, due to whatever reasons, has been from Indians rather than outsiders.

I do not want to sound too critical but one of the greatest problems with our culture is that we do not accept the treasures that lie with us, and when the same treasures are imported from the west we embrace them with open arms.

When Yoga had prominence only in India, not many people felt the desire to practice it. But when the western world and the celebrities of the west started showing interest in Yoga even we wanted to learn it.

This is not a recent phenomenon. This has happened with almost all of our spiritual and ancient sciences. The philosophy of the Buddha has been accepted more by the other countries of Asia than India. Ancient practices of Zen religion were not readily accepted by people here, but when an American author wrote a book by the name of ZEN *and the Art of Motorcycle Maintenance*, it became a bestseller in India!

At this juncture, I am precisely concerned about the same thing. Vedic Mathematics is getting a red-carpet welcome from the educational circles in the west whilst many people in its country of origin are not much interested in teaching and making it popular.

As the author of this book and as one of the world's youngest faculties on this subject, it is my sincerest desire that we realize the value and greatness of a science which has been discovered by one of our fellow countrymen.

Otherwise, as it has always happened with ancient sciences, the west will realize its importance much ahead of us and put it to practice, and we shall only welcome it many years later through their channels....

Appendices

APPENDIX A

Technique for multiplying five-digit numbers

(a) * * * * *
     * * * * *

(b) * * * * *
     * * * * *

(c) * * * * *
     * * * * *

(d) * * * * *
     * * * * *

(e) * * * * *
     * * * * *

(f) * * * * *
     * * * * *

(g) * * * * *
     * * * * *

(h) * * * * *
     * * * * *

(i) * * * * *
     * * * * *

APPENDIX B

In this appendix, we will study how the Criss-Cross system is used to multiply algebraic identities. For the steps used in the multiplication procedure refer to the chapter on Criss-Cross multiplication.

Q. Multiply $(a + b)$ by $(a + 3b)$

$$a + b$$
$$\underline{a + 3b}$$

- We multiply the right hand most terms $(b \times 3b)$ and write the answer as $3b^2$.

  (Answer at this stage is _____ $3b^2$)

- Next, we multiply $(a \times 3b) + (a \times b)$. The answer is 3ab plus ab, which equals 4ab.

  (Answer at this stage is _____ $4ab + 3b^2$)

- Finally, we multiply the left most terms $(a \times a)$ and write the answer as a^2.

  (Final answer is $a^2 + 4ab + 3b^2$)

Other examples:

(a) $2a + 5b$ (b) $5m^2 + 5m + 5$
     $\underline{3a + 5b}$ $\underline{5m^2 + 5m + 5}$
     $6a^2 + 25ab + 25b^2$ $25m^4 + 50m^3 + 75m^2 + 50m + 25$

Sometimes it may happen that a particular power of x is absent in the multiplication. In such a case, it should be given a zero value and the multiplication should be continued in the same manner.

Q. Multiply $(2x^3 + x + 5)$ by $(3x^3 + 5x^2 + 1)$

In the first bracket the term x^2 is absent and in the second bracket the term x is absent. We will take their values as zero and solve the problem

$$2x^3 + 0x^2 + x + 5$$
$$3x^3 + 5x^2 + 0x + 1$$
$$\overline{6x^6 + 10x^5 + 3x^4 + 22x^3 + 25x^2 + x + 5}$$

APPENDIX C

In the chapter 'Dates & Calendars' I mentioned Zeller's Rule, which is used to calculate the day on which any date falls for any year. With this technique you will have the calendar for any given year available to you. Zeller's Rule has no practical application for students but still some of you might be eager to learn the technique and hence I have included it in the Appendix. Since the rule is standardized, I am keeping the variables exactly as they are.

You can find the day on which any date falls using the formula:

$$F = k + [(13 \times m\text{-}1)/5] + D + [D/4] + [C/4] - 2 \times C$$

k – Date

m – Month number

D – Last two digits of the year

C – The first two digits of the century

RULES

- In Zeller's Rule the year begins with March and ends in February. Hence, the month number for March is 1, April is 2, May is 3 and so on up to January, which is 11, and February is 12.

- January and February are counted as the 11th and 12th months of the previous year. Hence, if we are calculating the day of any date on January 2026, the notation will be (month = 11 and year = 25) instead of (month = 1 and year = 26).

- While calculating, we drop off every number after the decimal point.
- Once we have found the answer we divide it by 7 and take the remainder. Remainder 0 corresponds to Sunday; remainder 1 corresponds to Monday; remainder two corresponds to Tuesday; and so on....
- If the remainder is negative, then add seven.

Examples:

Q. Find the day on 26th June 1983

(Here, k is 26, m is 4, D is 83, and C is 19.)

$$F = k + [(13 \times m - 1)/5] + D + [D/4] + [C/4] - 2 \times C$$
$$= 26 + [(13 \times 4 - 1)/5] + 83 + [83/4] + [19/4] - 2 \times 19$$
$$= 26 + [51/5] + 83 + [20.75] + [4.75] - 38$$
$$= 26 + 10 + 83 + 20 + 4 - 38 \text{ (we drop the digits after the decimal)}$$
$$= 105$$

When 105 is divided by 7, the remainder is 0 and hence the day is a Sunday. Thus, 26th June 1983 is a Sunday.

Q. Find the day on 4th February 2032

As mentioned in the rules, February is taken as month number 12 and the year will be the previous year. Hence, the value of D will be 31 instead of 32.

(The values of k, m, D, and C are 4, 12, 31 and 20 respectively.)

$$F = k + [(13 \times m - 1)/5] + D + [D/4] + [C/4] - 2 \times C$$
$$= 4 + [(13 \times 12 - 1)/5] + 31 + [31/4] + [20/4] - 2 \times 20$$
$$= 4 + [155/5] + 31 + 7 + 5 - 40$$
$$= 4 + 31 + 31 + 7 + 5 - 40$$
$$= 38$$

When 38 is divided by 7 the remainder is 3 and hence the day is a Wednesday.

APPENDIX D

Pythagorean Values

The Theorem of Pythagoras is one of the most important theorems in geometry. Right from high school students to those appearing for competitive exams to engineers, everybody deals with this theorem as a part of their daily practice. The concept that we will study in this Appendix is not only useful for calculating Pythagorean triplets but also useful in the Apollonius theorem, co-ordinate geometry and trigonometric identities like $\sin^2\varnothing + \cos^2\varnothing = 1$.

We know that the square of the hypotenuse of a right-angled triangle is equal to the sum of the squares of the other two sides. For example, if the sides of the right-angled triangle are 3, 4 and 5 then the square of 5 equals the square of 3 plus the square of 4. The values 3, 4 and 5 together form a Pythagorean triplet. Other examples of triplets are (6-8-10), (9-40-41), (11-60-61), etc. We are going to study a method by which we can form triplets out of a given value. This technique will be of great help to students dealing with geometric figures in competitive exams.

We will divide our study into two parts. In the first part, we will express the square of a number as the sum of two squared numbers. In the second part, we will express a given number as the difference of two squared numbers. The first part is further divided into two cases — the first case dealing with odd numbers and the second dealing with even numbers.

(A) Expressing the square of a number as the sum of two squared numbers

Case 1: Odd Numbers

Rule: The square of an odd number is also odd. This square is the sum of two consecutive middle digits.

Example: (a) $3^2 = 9 = 4 + 5$

(b) $5^2 = 25 = 12 + 13$

(c) $9^2 = 81 = 40 + 41$

We see that the square of 3 (odd) is 9 (odd), which is the sum of two consecutive numbers, viz. 4 and 5. Similarly, 25 is the sum of two consecutive numbers, 12 and 13.

But how do we find the consecutive numbers? The answer is very simple. To find the consecutive numbers, just divide the square by 2 and round it off.

In the first example, the square of 3 is 9. When 9 is divided by 2 the answer is 4.5. When 4.5 is rounded off to its next higher and lower numbers we have 4 and 5! Thus, (3-4-5) form a triplet where the square of the highest number is equal to the sum of the square of the other two numbers.

In the second example, the square of 5 is 25. When 25 is divided by 2 we get 12.5. When 12.5 is rounded off to its next higher and lower numbers, we have 12 and 13. Thus, (5-12-13) form a triplet where the square of 13 equals the square of 5 plus the square of 12.

In the last example, the square of 9 is 81. When 81 is divided by 2 and rounded off it gives 40 and 41. Thus, (9-40-41) is a triplet where the square of the biggest number (41) is equal to the sum of the square of the other two numbers (9 and 40).

Similarly, $35^2 = 1225$ (1225/2 = 612.5)

Thus, 35, 612 and 613 form a triplet where 613 units is the length of the hypotenuse and 35 units and 612 units are the lengths of the other two sides of the triangle.

Case 2: Even Numbers

The square of an even number is an even number. Therefore, we cannot have two middle digits on dividing it by 2. Thus, we divide the even number by 2, 4, 8, 16, etc. (powers of 2) until we get an odd number. Once we get an odd number we continue the procedure as done in the case of odd numbers.

If you have divided the even number by 2, 4, 8, etc. the final answer will have to be multiplied by the same number to get the triplet.

Q. One value of a Pythagorean triplet is 6; find the other two values.

We divide 6 by 2 to get an odd number 3. Next, we follow the same procedure as explained in the case of odd numbers to form a triplet (3-4-5). Now, since we have divided the number by 2 we multiply all the values of (3-4-5) by 2 to form the triplet (6-8-10).

Thus, the other two values are 8 and 10.

Q. One value of a triplet is 20; find the other two values.

We divide 20 by 4 to get an odd number 5. Next, we follow the same procedure as explained before to form a triplet (5-12-13). Now, since we have divided the number by 4, we multiply all values of the triplet (5-12-13) by 4 and make it (20-48-52).

Thus the other two values are 48 and 52

(B) Expressing a given number as a difference of two squares

We express a given number 'n' as a product of two numbers 'a' and 'b' and then express it as:

$$n = [(a + b)/2]^2 - [(a - b)/2]^2$$

Q. Express 15 as a difference of two squared numbers.

We know that $15 = 5 \times 3$. Thus, the value of 'a' is 5 and 'b' is 3.

Thus, $15 = [(5 + 3)/2]^2 - [(5 - 3)/2]^2$
$$= (8/2)^2 - (2/2)^2$$
$$= (4)^2 - (1)^2$$

(We can verify that $16 - 1 = 15$)

From the above, it can be proved that if one side of the triangle is $\sqrt{15}$ then the other side is 1 unit and the hypotenuse is 4 units. This can be checked with the expression $(\sqrt{15})^2 + 1^2 = 4^2$.

Q. In $\triangle ABC$, the value of side angle B is 90 degrees and side BC is 6 units. Find a few possible values of side AB and side AC.

Since the triangle is a right angled triangle, $BC^2 = AC^2 - AB^2$. The value of $BC^2 = 36$. Thus, $36 = AC^2 - AB^2$. In other words, we have to express 36 as a difference of two squares. We will apply the same rule that we used in the previous example.

We have to express the number 36 as a product of 2 numbers 'a' and 'b'.

(a) $36 = 18 \times 2$ (b) $36 = 9 \times 4$ (c) $36 = 12 \times 3$
    $= [(18+2)/2]^2 - [(18-2)/2]^2$ $= (13/2)^2 - (5/2)^2$ $= (15/2)^2 - (9/2)^2$
    $= (20/2)^2 - (16/2)^2$ $= (6.5)^2 - (2.5)^2$ $= (7.5)^2 - (4.5)^2$
    $= 10^2 - 8^2$ $= 6.5, 2.5$ $= 7.5, 4.5$
    $= 10, 8$

In example (a) the value of sides AC and AB is 10 units and 8 units.

In example (b) the value of sides AC and AB is 6.5 units and 2.5 units.

In example (c) the value of sides AC and AB is 7.5 units and 4.5 units.

We can verify:

$$6^2 = 10^2 - 8^2$$
$$6^2 = 6.5^2 - 2.5^2$$
$$6^2 = 7.5^2 - 4.5^2$$

From the examples we can conclude that if the length of side BC is 6 units, then the lengths of sides AC and AB can be either 10 and 8 or 6.5 and 2.5 or 7.5 and 4.5 units.

We have thus formed three triplets (6,8,10) (2.5, 6, 6.5) and (4.5, 6, 7.5).

APPENDIX E

Divisibility Tests

Given below are the divisibility tests of numbers. I have also included the divisibility test of numbers like 7, 11 and 13 which are not so popular.

Divisible by	Condition
3	Add up the digits. If the sum is divisible by 3 then the number is divisible by 3.
4	If the number formed by the last 2 digits is divisible by 4, then the whole number is divisible by 4.
5	If the last digit is either 5 or 0
6	Check for divisibility tests of 2 and 3. If divisible by 2 and 3 then it is divisible by 6.
7	Double the last digit and subtract it from the remaining number. If what is left is divisibly by 7, then the original number is also divisible.

Divisible by	Condition
8	If the last three digits are divisible by 8, then the whole number is divisible.
9	Add the digits. If the sum is divisible by 9 then the whole number is divisible by 9. (This holds true for any power of three)
10	If the number ends in 0
11	If the difference between • the sum of 1st, 3rd, 5th digits…..and • the sum of 2nd, 4th, 6th digits is a multiple of 11 or 0
12	Check for divisibility by 3 and 4
13	Delete the last digit from the given number. Then, subtract 9 times the deleted digit from the remaining number. If what is left is divisible by 13, then so is the original number.

In the same way we can continue the table….For divisibility by 14, check for divisibility by 2 and 7. For divisibility by 15, check for divisibility by 3 and 5 and so on…

APPENDIX F

Raising to fourth and higher powers

We have studied the technique of calculating cubes in the 'Advance Level.' The same technique can be used for calculating the fourth and higher powers. We will extend the logic used for cubing numbers and apply it for calculating the fourth and higher powers.

$$(a+b)^4 = a^4 + 4a^3b + 6a^2b^2 + 4ab^3 + b^4$$

We can represent it as:

$$a^4 + a^3b + \quad a^2b^2 + \quad ab^3 + b^4$$
$$+ \quad 3a^3b + 5a^2b^2 + 3ab^3$$

which on addition gives $(a + b)^4$

(Q) What is the answer of 21 raised to the power of 4

In this case, $a = 2$ and $b = 1$

The terms are in the ratio of b/a as we move from left to right and a/b if we move from right to left (as seen in cubing).

Steps:

- In the first row, we will find the fourth power of 2 and subsequently multiply it with the ratio ½ till we get the five terms.
- We will then multiply the second term with 3, the third term with 5 and the fourth term with 3 and write it as the second row.
- After adding the terms, we will put 4, 3, 2, 1 and no zeros behind the terms and add them once again.

$(21)^4 =$	16	8	4	2	1
		24	20	6	
	16	32	24	8	1

$$= 160000$$
$$32000$$
$$2400$$
$$80$$
$$1$$
$$\overline{194481}$$

Ans: When 21 is raised to the power of 4 the answer is 194481.

In the same way, we can find the fourth power of any number. We can further expand the logic to find the fifth and higher powers too.

APPENDIX G

Co-ordinate Geometry

The Sutras of Vedic Mathematics cover the topics of Geometry too. Let us have a look at an example from co-ordinate geometry. We will find the equation of a straight line passing through two points whose co-ordinates are known.

(Q) Find the equation of a straight line passing through the points (7, 5) and (2, -8).

There are two approaches of solving the question through the traditional method. The first approach is using the formula $y = mx + c$. The second approach is using the formula:

$$y - y_1 = \frac{y_2 - y_1}{x_2 - x_1} (x - x_1)$$

Traditional Method One:

In this method we take the general equation as $y = mx + c$. On substituting the above values, we have: $7m + c = 5$; and $2m + c = -8$.

Next, we solve the equations simultaneously:

$$7m + c = 5 \ldots\ldots\ldots\ldots\ldots(1)$$
$$\underline{-2m + c = -8 \ldots\ldots\ldots\ldots\ldots(2)}$$

Therefore, $5m = 13$; $\ m = \dfrac{13}{5}$

We now substitute the value of 'm' in a equation (1), We have,

$$7 \times \frac{13}{5} + c = 5 \rightarrow \frac{91}{5} + c = 5 \rightarrow c = 5 - \frac{91}{5} \rightarrow c = \frac{-66}{5}$$

Substituting the values of m and c in the original equation (y = mx + c), we have;

$y = \frac{13x}{5} - \frac{66}{5}$. And therefore, we have the equation of the line as **13x - 5y = 66.**

We have seen the first traditional method of solving the problem. Now, let us have a look at the second traditional method using the formula $y - y_1 = \frac{y_2 - y_1}{x_2 - x_1} (x - x_1)$

Traditional Method Two:

On substituting the values (7, 5) and (2, -8), we have:

$$y - 5 = \frac{-8 - 5}{2 - 7} (x - 7)$$

$$= (y - 5) = \frac{-13}{-5} (x - 7)$$

$$= -5(y - 5) = -13 (x - 7)$$

$$= -5y + 25 = -13x + 91$$

$$13x - 5y = 66$$

We can use any of the two methods mentioned above to find the equation of the straight line. However, these methods are lengthy, tiring and cumbersome. Plus there are high chances of making errors in working the complicated steps.

Vedic Mathematics, on the other hand, offers such a simple solution that you will be amazed to see how easy it is to get the answer. The Vedic mental method of solving such problems is as follows:

'Put the difference of y co-ordinates as the x co-efficient and the put the difference of x co-ordinates as y co-efficient'

Vedic Mathematics Method:

The given co-ordinates are (7, 5) and (2, -8)

Therefore, our x co-efficient is 5-(-8) = 13 and

our y co-efficient is 7-2   =   5.

We have the answers 13 and 5 with us. Thus, the LHS is $13x - 5y$.

Next, the RHS can be easily obtained by substituting the values of x and y of any co-ordinate in the LHS. For example, $13(7) - 5(5) = 66$. Thus, our final equation becomes **$13x - 5y = 66$**.

An alternate way of obtaining the RHS is using the rule:

'Product of the means minus the product of the extremes'

Therefore, we have (7, 5) and (2, -8)

$$= (5 \times 2) - (-8 \times 7)$$
$$= 66$$

Thus, the technique can be easily verified !

One can see from the examples mentioned above, the simplicity and straight-forwardness of the Vedic Mathematics approach. Problems on geometry which take 5-7 steps can be solved mentally just by looking at them!

Answers

Chapter 1

1	(a) 2025	1	(b) 9025	1	(c) 13209	1	(d) 11021
2	(a) 3136	2	(b) 2601	2	(c) 2809		
3	(a) 566433	3	(b) 2324876751	3	(c) 659934	3	(d) 30199698
4	(a) 354431	4	(b) 710699	4	(c) 1334331	4	(d) 8884551
5	(a) 539385	5	(b) 27148	5	(c) 493284	5	(d) 400599
6	(a) 324.57	6	(b) 2390.02	6	(c) 9334	6	(d) 997.347

Chapter 2

A (1) 276	A (2) 374	A (3) 693	A (4) 533
A (5) 67520	A (6) 24642	A (7) 63630	A (8) 1234321
B (1) 968	B (2) 1353	B (3) 2821	B (4) 1224
B (5) 39738	B (6) 85446	B (7) 481401	B (8) 7507936
C (1) 2176	C (2) 1435	C (3) 1612	C (4) 7462
C (5) 157899	C (6) 390819	C (7) 268932	C (8) 39325421

Chapter 3

A (1) 1764	A (2) 1089	A (3) 13225	
B (1) 42025	B (2) 4020025	B (3) 16402500	B (4) 99820081
B (5) 96040000	B (6) 1188100		
C (1) 6724	C (2) 2401	C (3) 11881	C (4) 9409

Chapter 4

A (1) 99	A (2) 87	A (3) 68	A (4) 48
A (5) 36	A (6) 18	A (7) 75	A (8) 101
B (1) 51	B (2) 62	B (3) 63	B (4) 98
B (5) 78	B (6) 29	B (7) 32	B (8) 21

Chapter 5

A (1) 96   A (2) 87   A (3) 73   A (4) 58
A (5) 41
B (1) 99   B (2) 75   B (3) 44   B (4) 59
B (5) 37
C (1) 113   C (2) 125   C (3) 152   C (4) 109

Chapter 6

1 (a) 984060   1 (b) 999992000007 1 (c) 10120200
1 (d) 1055250

2 (a) 272   2 (b) 1560000   2 (c) 88350000 2 (d) 857420

3 (a) 9984   3 (b) 893560   3 (c) 100396800 3 (d) 979700

4 (a) 72781   4 (b) 93060   4 (c) 819590   4 (d) 103020

5 (a) 2352   5 (b) 484   5 (c) 2597   5 (d) 306

5 (e) 247504

Chapter 7

A (1) 49   A (2) 9025   A (3) 972196   A (4) 1050625
A (5) 1024144
B (1) 7225   B (2) 774400   B (3) 828100   B (4) 324
B (5) 1254400
C (1) 484   C (2) 41209   C (3) 91809   C (4) 248004
C (5) 50625

Chapter 8

A (1) 8   A (2) 2   A (3) 5   A (4) 0/9
B (1) Correct   B (2) Correct   B (3) Incorrect   B (4) Correct
B (5) Incorrect   B (6) Correct   B (7) Correct   B (8) Correct
B (9) Correct
C (1) Alternative 3   C (2) Alternative 4

Chapter 9

A (a)

8	6	16
18	10	2
4	14	12

A (b) Yes
A (c) 10
A (d) 30
B (a) 58

33	30	12	69	51
54	36	18	15	72
75	57	39	21	3
6	63	60	42	24
27	9	66	48	45

(b) The sides will give equal totals by themselves

(c) The value of centre-most square will be 12 units more in case (b) as compared to case (a).

(d) The difference in final total will be 60 units (12 × 5)

C (a)

20	15	40
45	25	5
10	35	30

40	5	30
15	25	35
20	45	10

30	35	10
5	25	45
40	15	20

10	45	20
35	25	15
30	5	40

C (b)

22	21	13	5	46	38	30
31	23	15	14	6	47	39
40	32	24	16	8	7	48
49	41	33	25	17	9	1
2	43	42	34	26	18	10
11	3	44	36	35	27	19
20	12	4	45	37	29	28

The first son should receive the cows as shown in the first row. The second son should receive the cows as shown in second row and so on...

Chapter 10

A (1) Leap Years: 2000, 2040, 2004, 1004, 2404, 1404, 4404
    Not Leap Years: 2100, 2101

A (2) (i) Friday   (ii) Saturday   (iii) Monday   (iv) Sunday
    (v) Sunday   (vi) Friday

B (1) Tuesday   B (2) Friday   B (3) Tuesday   B (4) Thursday

C (1) Thursday   C (2) Friday   C (3) Tuesday   C (4) Friday

Chapter 12

A (1) 4,3   A (2) 5,2   A (3) 0.5, 0.5   A (4) 1, 6

A (5) 4.5, 2   A (6) 5, 10

B (1) 0,2   B (2) 4,0

C (1) The number of one rupee and two rupee coins are 23 and 29 respectively

$(x + y = 52, x + 2y = 81)$

C (2) The monthly income of Tom is Rs. 3200

$(4x - 3y = 800, 3x - 2y = 800)$

C (3) The numbers are 54 and 36. (In this case the first equation is $(x + y)/2 = 90$ and therefore $x + y = 180$. The second equation is $2x = 3y$.)

C (4) The number of students in each classroom are 60 and 80 respectively. The first equation is $\dfrac{x - 10}{y + 10} = \dfrac{5}{9}$. The second equation is $\dfrac{x + 10}{y - 10} = \dfrac{1}{1}$.

Chapter 13

A (1) 31   A (2) 83   A (3) 111   A (4) 2021

B (1) 26.45   B (2) 39.37   B (3) 125.06

C (1) 0.63   C (2) 12.26

Chapter 14

A (1) 1157625   A (2) 68921   A (3) 157464   A (4) 12167

A (5) 117649   A (6) 7290   A (7) 7880599   A (8) 884736

B (1) 287496   B (2) 456533   B (3) 753571   B (4) 6859

C (1) 79507   C (2) 373248   C (3) 1030301

Chapter 15

A (1) Q = 1; R = 28   A (2) Q = 11, R = 235   A (3) Q = 12; R = 6042

A (4) Q = 10; R = 15

B (1) Q = 4; R = 39   B (2) Q = 106; R = 19   B (3) Q = 13; R = 443

B (4) Q = 12; R = 3255   B (5) Q = 346; R = 9

Chapter 16

A (a) Q = 12; R = 33   A (b) Q = 110, R = 206   A (c) Q = 11, R = 3348

A (d) Q = 2200; R = 6

B (a) Q = 119; R = 26   B (b) Q = 20; R = 830   B (c) Q = 109; R = 71

B (d) Q = 125; R = 156

C (a) Q = 8; R = 141   C (b) Q = 30; R = 39

Frequently Asked Questions

FACT FILE

Name	Vedic Mathematics
Founder	Jagadguru Swami Sri Bharati Krishna Tirthaji Maharaja
Country	India
Year	(around) 1957
Contents	16 word-formulae and some sub-formulae
Uses	• Quick Calculation • Solving math problems mentally

Q. What is Vedic Mathematics?

Vedic Mathematics is the name given to a set of word-formulae as invented by Jagadguru Bharati Krishna Maharaj. The formulae contain secrets of quick calculation and solving mathematical problems mentally.

Q. Is it taken from the Vedas?

As such, the science is not taken from the Vedas. It is an original endeavour of the author. However, the appendices to the Atharvaveda contain some parallel references.

Q. What is the best age to start teaching it?

The techniques of Vedic Mathematics are not accepted by the normal school systems. Using Vedic Mathematics, the answer to a particular problem is obtained directly without any intermediary steps. However, students are given marks for showing the intermediary steps and the calculation procedure involved in solving a particular problem. Hence, any particular Vedic technique should be taught to a student only after he has mastered the traditional way (the method taught in school).

Example: A third standard student is expected to show the complete steps while solving a multiplication problem. The teacher will give him full marks after carefully observing all the steps. However, a fifth or a sixth grade student is not required to show all the steps while solving a multiplication problem. If his final answer is correct, the teacher may well award him full marks. Thus, there is no ideal age as such. However, most concepts of Vedic Mathematics will be best understood by high school and college students.

Q. How is VM different from the abacus technique of calculation?

The abacus system of calculation is Chinese whilst VM is Indian. Secondly, the abacus system deals with arithmetic only and hence it is taught to students of the age group of 5-13 years. On the other hand, VM not only covers arithmetic but also algebra, geometry, calculus and other advanced branches of mathematics. Hence, it is suitable for students of any age.

Q. What is the international response to VM?

VM has received a tremendous response from the United Kingdom, USA, Poland, Ireland, Germany, Italy, South Africa and many other countries all across the world. A centre for VM was established in Singapore in 1999 where Kenneth Williams was invited from the UK to prepare the curriculum. The

London School of Economics has introduced VM in its affiliated
St. James Independent Schools in the UK. Experts like James
Glover and Jeremy Pickles have written many books and
research articles on the subject. W. Bradstreet Stewart of
Sacred Science Institute, USA, has done considerable research
in recent years.

Q. How do we attend/organise seminars on VM?

To attend/organise seminars and training programs in your
city, contact the author on dhaval@dhavalbathia.com

Bibliography

QUICK CALCULATION

- *Vedic Mathematics* by Jagadguru Swami Sri Bharati Krishna Tirthaji Maharaj
- *Speed Mathematics* by Bill Hanley
- *The Trachtenberg Speed System of Basic Arithmetic* by Ann Cutler and Rudolph McShane
- *Vedic Mathematics for Schools* (Part 1, 2 and 3) by James Glover

EFFECTIVE STUDIES

- *The Portrait of a Super Student* by Abhishek Thakore
- *How To Top Exams and Enjoy Studies* by Dhaval Bathia
- *Study Smarter Not Harder* by Kevin Paul
- *How To Pass Exams Without Anxiety* by David Acres

MIND-POWER SCIENCES

- *Mind Maps* by Tony Buzan
- *How To Develop A Super Memory* by Harry Lorayne
- *The Miracle of Mind Power* by Dan Custer
- *The Silva Mind Control Method* by Jose Silva

FEEDBACK FORM

(A) Would you like to receive more information about books, websites and newsletters on Vedic Mathematics and other subjects related to competitive exams like CAT, CET, IAS, IPS, IIT, MCA, etc.?

(B) Are you a parent who seeks more information on books and courses on self-development for your children?

(C) Would you be interested in knowing about the seminars/ training programs that happen in your city on Vedic Mathematics, memory systems and other mind-power sciences (especially for students giving board exams)?

(D) Are you a faculty associated with any coaching class/ institution or possess some material useful for students in professional and competitive exams?

(E) Are you a Principal, Head, Trustee or Professor of any school, college or coaching class and keen on teaching your students some powerful techniques that will help them improve their performance in exams?

(F) Are you associated with any NGO, club, social group, religious or charitable organisation and interested in conducting various training programs for your members?

(G) Are you an editor/journalist of any newspaper/magazine and want to publish any articles (or start columns) on Vedic Mathematics, mind-power sciences, or techniques to excel in exams, etc.?

If your answer is 'Yes' to any of the above questions then fill in the form given on the next page and send it to the author at the address mentioned below. You will be added to the author's mailing list and database and you will receive the necessary information. If you want more copies of the form, you can photocopy them for your friends and relatives.

(NOTE: Seven categories are mentioned above with the letters A, B, C, D, E, F and G. Please tick one or more categories which interest you in the form given in the next page. You can tick as many categories as you want.)

Send your forms to:

Dhaval Bathia,
40 Deepak Niwas,
Bhogilal Phadia Road,
Opp. S.V.P. School
Kandivali (w)
Mumbai 400067
Maharashtra, India

(E-Mail: dhaval@dhavalbathia.com)

(Kindly send more details about yourself, your organization, the seminars that you are keen on attending/organizing and other relevant details on separate sheets of paper along with this form. If you live abroad then please send an additional copy of the details via E-mail.)

Name: _____

Address: _____

_____

Pin: _____ State: _____ Country: _____

Tel: (1)_____ (2) _____ (with STD/ISD)

E-Mail: _____

Occupation: _____ (if student, write student)

Institution/School/College: _____

Categories in which you are interested (refer to previous page and tick the appropriate categories):

(A)___ (B)___ (C)___ (D)___ (E)___ (F)___ (G)___

Feedback (about this book): _____

_____

_____

_____

_____

_____

_____

Any other queries/comments: _____

_____

_____